Fundamentals
of
Floating
Production Systems

Fundamentals
of
Floating
Production Systems

Niladri Kumar Mitra

ALLIED PUBLISHERS PVT. LTD.

New Delhi • Mumbai • Kolkata • Lucknow • Chennai
Nagpur • Bangalore • Hyderabad • Ahmedabad

ALLIED PUBLISHERS PRIVATE LIMITED

Regd. Off. : 15 J.N. Heredia Marg, Ballard Estate, Mumbai–400001, Ph.: 022-22626476
E-mail: mumbai.books@alliedpublishers.com

12 Prem Nagar, Ashok Marg, Opp. Indira Bhawan, Lucknow–226001, Ph.: 0522-2614253
E-mail: appltdlko@sify.com

Prarthna Flats (2nd Floor), Navrangpura, Ahmedabad–380009, Ph.: 079-26465916
E-mail: ahmbd.books@alliedpublishers.com

3-2-844/6 & 7 Kachiguda Station Road, Hyderabad–500027, Ph.: 040-24619079
E-mail: hyd.books@alliedpublishers.com

5th Main Road, Gandhinagar, Bangalore–560009, Ph.: 080-22262081
E-mail: bngl.books@alliedpublishers.com

1/13-14 Asaf Ali Road, New Delhi–110002, Ph.: 011-23239001
E-mail: delhi.books@alliedpublishers.com

17 Chittaranjan Avenue, Kolkata–700072, Ph.: 033-22129618
E-mail: cal.books@alliedpublishers.com

81 Hill Road, Ramnagar, Nagpur–440010, Ph.: 0712-2521122
E-mail: ngp.books@alliedpublishers.com

751 Anna Salai, Chennai–600002, Ph.: 044-28523938
E-mail: chennai.books@alliedpublishers.com

Website: www.alliedpublishers.com
© 2009, Allied Publishers Pvt. Ltd.

ISBN 13: 978-81-8424-389-5

Published by Sunil Sachdev and printed by Ravi Sachdev at Allied Publishers Pvt. Ltd. (Printing Division), A-104 Mayapuri Phase II, New Delhi-110064

Acknowledgements

At the very outset, I would like to thank all the persons, personalities and events I have come across in my professional and academic life spanning over four decades that have shaped my thoughts and perceptions over a whole range of issues pertaining to our industry. I have been immensely benefited from the discussions, deliberations, brainstorming and mentoring exercises undertaken with my colleagues, peers, seniors and industry partners over these years. Also, my participations in various international and national conferences, seminars and workshops across the world and my visits abroad to a number of fields and design centers have infused in me the best of learning through the ensuing discussions thereof. I take this opportunity to thank all those events, discussions, materials and information which have amply helped me in organizing and substantiating my thoughts of this book for undergraduate students thereby bringing relevant industry information to the academics.

I would like to thank Shri M.S. Srinivasan the then Secretary, Petroleum and Natural Gas, Ministry of Petroleum and Natural Gas, Government of India and Shri R.S. Sharma, Chairman and Managing Director, ONGC groups of companies, for consistently inspiring, motivating and guiding me to write down the book incorporating my offshore field experiences as well as incorporating the experiences gathered through various interactions and deliberations over floating production systems for Indian offshore waters. While discussing industry-academy knowledge interface, Shri Srinivasan and Shri Sharma, lamented the lack of basic simple book available on floating production systems and wished that as oil and gas future belongs to deep waters and marginal fields, the student and the new incumbent to industry must have a basic fundamental knowledge of floating production system and sub-sea production system. Taking inspiration from them, I took the onus on me and the result is before you. I thank Shri Shrinivasan and Shri Sharma for reposing faith in me for undertaking such an exercise.

My heart-felt thanks goes to Shri Ashok Kumar Suptdg. Engr. (Prod.) of ONGC for volunteering himself, coming forward and doing his bit in discussing and deliberating the concepts with me, jotting down my experiences, soliciting my views and thereafter putting those guidance, substances and understandings into right words, right context and in right perceptions. I thank him for his passionate efforts towards comprehending, drafting, re-drafting and finalizing the book quite comprehensively. I would also like to thank Mr. Amarjyoti Das Suptdg. Engr. (Prod.) of ONGC, for doing all the groundwork with respect to the printing and publishing of the book. It was his tireless efforts with the Allied Publishers that has paved the way for this book coming to the market.

I would like to thank M/s Allied Publishers to come forward and partnering with us to print and publish technical and specialized book of this nature. The effort of M/s Allied is praiseworthy without which this book wouldn't have reached the campuses and other learning centers, which to my understanding, is the need of the hour so as to ensure a knowledge sharing amongst the new incumbents of this industry.

Niladri Kumar Mitra
ONGC, New Delhi, India

M.S. SRINIVASAN
Secretary

भारत सरकार
पेट्रोलियम एवं प्राकृतिक गैस मंत्रालय
शास्त्री भवन नई दिल्ली--११०००१

Government of India
Ministry of Petroleum & Natural Gas
Shastri Bhawan, New Delhi - 110 001

Foreword

The ever increasing demand for oil & gas and spiraling oil prices have put a sharp focus on the early development and monetization of new & small marginal fields as well as forays into the deep waters, as the existing fields all over the world are past the peak oil phase and are in the matured phase warranting substantial investments to produce at substantial higher rates. Economic considerations advocate the use of floating production systems for early production from these marginal fields, from the deep waters and from the existing ageing fields under redevelopment. This necessitates a good understanding of floating production systems in their given perspective by the undergraduate students about to join the field and also by those torch-bearers of the industry having no experience/exposures so far over floating production systems.

Though a number of books and journals are available on floating production systems, the need for a single comprehensive book written in a lucid and comprehensive way so as to serve as a basic reference book for under-graduate students was acutely felt. Available books are rather too complex to be understood by the fresher at the academics who has not seen the field and who does not have any inkling of operations & other critical trials, tribulations & stipulations of the high seas.

I am happy that Mr. Mitra has understood the need of the hour and authored this very fundamental book on floating production systems. And in the process of producing this book, he has put all his experiences in the given perspective to ensure that undergraduate students understand the concepts, underlying principles and applications of floating production systems in a simple yet comprehensive way. While going through the book, I find that he has taken care to avoid complex issues. His underlying objective clearly appears to be for the book to act as a source of basic learning from where the students can develop further understanding once they join the field.

I am confident that the book will meet its intended objective, and wish the author more books from his pen.

(M.S. Srinivasan)

Phone : 011-23383501, 011-23383562 Fax : 011-23383100 E-mail : srinivasan2k@nic.in

आर. एस. शर्मा
R. S. Sharma
अध्यक्षा एवं प्रबन्ध निदेशक
Chairman & Managing Director

ऑयल एण्ड नेचुरल गैस कॉरपोरेशन लि.
Oil and Natural Gas Corporation Ltd.

Chairman, ONGC Group of Companies

Foreword

Hydrocarbons continue to dominate the global energy basket ever since they were discovered for commercial use almost 150 years ago. With the crude oil price peaking to unprecedented levels, the 'peak oil' theory has been gaining attention world-wide. While serious activity for developing non-conventional and alternate sources of energy is being undertaken globally, renewed efforts are also afoot to enhance oil and gas production. It is obvious that oil will continue to dominate as a fuel source for the foreseeable future and hence appropriate technology will be the key to finding 'new' oil as also to exploit existing resources.

Deep and Ultra-Deep waters are today accessible due to technological advances and also on account of the prevailing high oil price regime. Exploitation of discoveries from these locations requires use of innovative solutions such as Floating Production Systems that are also used for exploiting Marginal Offshore fields and also as Early Production Systems (EPS). This book is therefore timely as it provides information and insights into these facilities that will be increasingly deployed in years to come.

While literature on these systems can be accessed from myriad resources, my colleague on the Board of Directors of ONGC, Mr. N.K. Mitra brings in rich perspectives honed from more than three decades of his hands-on oil field experience. His insight and ability to foresee operational issues has been of immense value to ONGC. On perusal of the book's contents, I find that he brings the same rigour, analysis and application to this endeavour as well.

The book is comprehensive in content that is presented in a simple and effective manner. I am confident it will serve its intended purpose of providing first reference to students and new entrants to the oil industry.

(R.S. Sharma)

जीवन भारती, 124 इन्दिरा चौक, नई दिल्ली 110 001 भारत दूरभाष – 91 11 2332 3402, 2331 5607 फैक्स + 91 11 2331 0553, 2331 3028 ई-मेल cmd@ongc.co.in
Jeevan Bharati, 124 Indira Chowk, New Delhi 110001 India Phone - 91 11 2332 3402, 2331 5607 Fax - 91 11 2331 0553, 2331 3028 e-mail cmd@ongc.co.in

आइ. एस. एम. विश्वविद्यालय, धनबाद
I.S.M. UNIVERSITY, DHANBAD

Established under Section 3 of the UGC Act 1956 vide Notification
No. F.11-4/67-U3, dated 18th September 1967, of the Government of India)

भारतीय खनि विद्यापीठ विश्वविद्यालय, धनबाद - 826004 झारखण्ड, भारत
INDIAN SCHOOL OF MINES UNIVERSITY, DHANBAD-826004, JHARKHAND, INDIA

प्रोफेसर तारकेश्वर कुमार
Prof. T. Kumar
निदेशक
Director

Foreword

Earlier to 1970 there has been very little knowledge of recovering hydrocarbons from offshore fields especially in India. Though the other aspects of Offshore Engineering were known but the petroleum operations in offshore water was totally a new concept. When the first offshore drilling was done in Aliabet Structure off-Cambay in Gujarat, India, expertise were drawn from outside India. Gradually of course engineers and technologists from within picked up the science of operation in this new environment.

As more attention is being given towards exploration and exploitation for hydrocarbon in offshore areas, absence of good literature for better understanding has been felt. There is hardly any book where someone can get all technical information related to offshore operation.

It is probably with this problem in mind Mr. N.K. Mitra who not only is heading offshore division of ONGC now but all throughout his career remained associated in this environment with this operation only has thought to pen his experience in the form of a text book which will serve students both undergraduate and postgraduate and also practicing engineers.

The book titled "Floating Production System" deals one of the aspects of the whole gamut of Offshore Production System but there is no doubt that till now any other book has dealt so elaborately about this comparatively new area, i.e. floating production system. The book starts with the fundamentals and ends with the specialized topics. Each chapter is so nicely dealt with that anybody will feel like reading it from the beginning to the end. At the end the reader will have the feeling that he has learnt something new which has added to his knowledge.

Undoubtedly this will form a text book for the students and a treasure of information for all the practicing engineers in this area. I am sure it will help in filling up the gap between what is known and what is conceived.

[Prof. T. Kumar]

TEXAS A&M

THE HAROLD VANCE DEPARTMENT OF
PETROLEUM ENGINEERING

Akhil Datta-Gupta
LeSuer Chair in Reservoir Management
Texas A&M University
College Station. Texas 77843-3116
979-847-9030
a.datta-gupta@pe.tamu.edu

Foreword

Over the past several decades, offshore oil and gas production has increasingly moved into deeper waters and farther into remote areas away from existing infrastructures. The higher oil price and demand have also led the industry towards exploiting smaller and marginal fields. All these have made Floating Production Systems (FPS) an integral part of modem offshore oil field development. Selection of an optimal FPS is governed by many technical and environmental factors. To my knowledge, this is one of the first books that systematically discuss the various aspects of FPS ranging from structures, facilities, sub-sea production systems, loading and transportation. The complex operations of the FPS are explained in a clear and simple language that makes it accessible to technical as well as non-technical audience. The book brings to bear over 30 years of offshore production experience by the author. Mr. Mitra is undoubtedly one of the stalwarts in offshore production and development in India. His contributions in deploying leading edge technology to revitalize and arrest production decline in Mumbai High, the largest offshore oil field in India, are well recognized. The book is an invaluable compilation of his unique perspective and insights into the offshore production in general and FPS in particular. I think the book is very topical and will fill in an important void in the petroleum engineering literature. The breadth of the topics covered here will be of great value to undergraduates and practicing engineers. The experience and insights provided by the author will go a long way in moving the technology forward.

Akhil Datta-Gupta
Professor and LeSuer Chair
Petroleum Engineering, Texas A&M University

3116 TAMV, College Station, Texas 77843-3116
(979) 845-2241/ FAX (979) 845-1307 / http://pumpjack.tamu.edu

Preface

Deep and ultra-deep waters and small and marginal fields currently holds the key for the oil security for any nation. Successful exploitation from these fields warrants applications of new thoughts and perspectives as well as new technological interventions. Floating Production System is one such concept to get early oil from these fields in most feasible way justifying economic considerations. Though FPSOs are in use since 1970's, its use has been increased substantially in recent times all over the world at varied water depths. Hence, it is incumbent upon those joining the industry or already serving the industry to have sound knowledge and understanding of Floating Production System, its related concepts and field development options through this in all its comprehensives so as to ensure better returns and better recovery from these fields in years to come.

Though a number of books and journals are available on Floating Production System, a single comprehensive book written in a simple but comprehensive way so as to serve as a basic fundamental book for undergraduate students was missing. This book is an attempt to fill this gap and provide the students a platform to develop a comprehensive and perspective understanding before they join the fields to ensure the needed deliverables.

The book "Fundamentals of Floating Production System" is a basic and fundamental book written for the students of undergraduate level. Things in the book have been put in easy language so as to be understandable by students in academics who have not seen the fields and don't have hands on experience of field's various deliberations and tribulations. Complexity has been avoided in the book deliberately and things are kept as simple and understandable as possible.

In the first chapter, various types of Floating Production System have been discussed in terms of their basic features, their associated peripherals, their advantages and disadvantages. At the very outset, the advantages of using a Floating Production System over the fixed platform structure, for the development of deep waters and small/marginal fields has been discussed so as to give a starting notion to the students over the relevance of Floating Production System in years to come. Further, through a logical flow diagram, efforts have been made to explain the underlying concepts of selection of any particular Floating Production System.

Mooring is an important aspect of any Floating Production System. Hence, Chapter 2 has been dedicated to understand various types of mooring systems available, their components, patterns and types of mooring along with their comparative advantages and disadvantages in the given Floating Production System as well as in the given field location perspectives.

Chapters 3, 4 and 5 give in comprehensive details the various process facilities, process support facilities, utilities systems, safety and firefighting systems available on any given Floating Production Systems. Actual field deployment may see addition or deletion of some more process and utility systems depending upon the field requirement and existing available facilities in the given field. These facilities and systems vary in terms of design, quantity and quality depending upon the field conditions and requirements. They also give recommendation for what types of facilities or rather say what combinations of facilities suits to different types of Floating Production Systems. Wherever little more understanding of some process or utility or safety or fire system is desired to have better understanding by students, the same has been done in these chapters for those systems. Further, some important concepts like that for sour gas fields and other relevant issues like fire and fire classification, hazardous area classifications, corrosion mechanisms etc have been discussed briefly in Annexure.

After, the students got themselves aware of the basics of the topside facilities on board of any Floating Production System in Chapters 3, 4 and 5, in this Chapter 6, the basic design and technical considerations of those different types of systems, process, utilities and other facilities has been discussed. Chapter gives a comprehensive list of various standards and codes that needs to be applied or considered while going for the design and technical consideration. Students are advised here to go through each of these codes and standards in more detail to understand the underlying facts and philosophies. In a very simple but in specific terms, chapter talks about the various initial or basic considerations like that of concept of modularization, effect of motion on process facilities, flow properties, corrosion considerations etc that has to be understood with respect to Floating Production Systems. Thereafter, the chapter deals with the space and layout considerations for equipment, encompassing important factors or considerations, at different Floating Production Systems as the layout varies from system to system.

Sub-sea production systems form a very important part for any Floating Production System so as to act as a comprehensive whole in a marginal field and deep sea ventures in a cost-effective perspective. Accordingly, Chapter 7 deals with the basics of various types of sub-sea production systems, their sub-systems/components and their major assemblies along with the little details in which they are specifically tailored to the specific field and operator's requirement.

Carrying forward the understandings of sub-sea production system of Chapter 7, Chapter 8 provides the various considerations and philosophies involved with the sub-sea production systems. Chapter talks about the basic differences between sub-sea production based field development and fixed platform based field development. Chapter deals with how the sub-sea production systems are configured and lay-out along with their comparative advantages and disadvantages. Chapter also talks in brief about design, operations and maintenance philosophy or statement for any equipment. This guideline policy, or philosophy or statement exists for all the equipment, either individually or for a group and hence, this understanding is important from design and engineering point of view.

Chapter 9 is an extension of Chapters 7 and 8 wherein the selection criteria of various types of sub-sea production system have been discussed for Barge-based, Tanker-based and Semi-submersible based Floating Production System. Here students can have a bird's eye view over how the selections procedures are decided and what are the components and criteria that need to be adopted for a given sub-sea production system associated with a given Floating Production System.

Chapter 10 deals with Offshore Crude Storage, Heating, Loading and Transportation systems wherein the offshore storage facilities, crude offloading systems and oil storage heating systems have been discussed in fundamental terms. To develop a better insight, chapter deals with the concepts of Tankers and Supertankers along with the basics of Tonnage measurement and Weight measurement. As most of the Floating Production Systems are basically converted supertankers, I thought it prudent to provide some details over tankers and supertankers to the students.

Finally Chapter 11 deals with a case study wherein a Field development scenario for a marginal field off Indian offshore shallow water has been considered with Floating Production Systems utilizing sub-sea production systems. Chapter needs to be understood along with the Figures provided in Tables 1 to 11. Figures in the chapter and in tables are just an approximation and has been put so as to comprehend and understand the concepts comprehensively. This chapter provides the student an idea of how field development scenarios are planned putting in perspective all the understandings made through the previous chapters. Only this comprehensive knowledge matters to the students once they join the field or the design centre. The deliverability of any knowledge over Floating Production System and Sub-sea production systems is always measured through how effectively, efficiently and optimally we comprehend or put across a field development scenario keeping the cost-objective as well as the physical and financial parameters in sharp-focus.

Some other important understanding and concepts like that of "Capital budgeting and Project financing", is required to be understood by the reader of this book. I would expect student to develop horizontal understanding of the related concepts in good comprehensiveness.

I have been to FPSO fabrication yard at Indonesia on few occasions wherein I have personally seen the things taking shape. Due discussions and deliberations undertaken thereof have played a key role in my understanding of the issues at the core of Floating Production Systems. I have incorporated those learning in the books at relevant places.

I hope that the basics and fundamentals as deliberated in various chapters of this book will give the students fairly good understanding of the Floating Production Systems and its attributes. I am sure that from here, they can go ahead with higher level of learning and understanding over each of those topics and sub-topics referring books, journals and through engaging in meaningful horizontal and vertical dialogues with field personnel and experts.

I would be happy to receive the suggestions and critical reviews of this book so as to incorporate those elements and those new perspectives in next editions. Together, we can bring out the best for the student.

N.K. Mitra
ONGC, New Delhi, India

Contents

CHAPTER 5: Floating Production System: Safety and 71
Fire Fighting System/Facilities

CHAPTER 6: Floating Production System: Process Facilities 96
and Utilities Design and Technical Considerations

CHAPTER 7: Sub-Sea Production Systems 109

Tables 199

Appendices 219

Glossary 235

Dedicated to

My Friends and Colleagues in the Industry
&
My wife Meenakshi, for standing by the side of this oil man through his thick and thin days in oil fields

"....... As we move deeper into offshore waters and as we wish an early production from deepwater, ultra-deepwater, remote, and marginal fields, in a cost-effective manner, floating production system coupled with sub-sea completion is the answer. Energy security in any emerging economy depends, to a large extent, how effectively we deploy these concepts to new water frontiers bearing substantial oil & gas reserves. Ever increasing oil prices, ever widening demand-supply gaps and economics of oil & gas development projects in recent times have made this floating production system & sub-sea production system much more relevant now. No doubt, future belongs to floating production system in-tandem with sub-sea production system ...."

Mitra N.K

Floating Production Systems: Introduction

Oil has been produced from offshore locations since the 1950s. Originally, all oil platforms sat on the seabed, but as exploration moved to deeper waters and more distant locations in the 1970s, the use of floating production systems got the prominence. It was in the late 1970s when the commercial use of Floating Production System started getting prominence owing to certain factors like advancement in sub-sea technologies, depressed oil prices, cost and schedule advantage of a MODU conversion besides other engaging factors during those periods. And currently too, this floating production system offers a much better solution for exploiting production from the marginal and the deepwater field developments owing to its versatility, its mobility in terms of relocation, its easy mobility and maneuverability during adverse weather conditions, its relatively low cost and also owing to its self-containment capabilities. About 100 floating production system deployed all over the world in its varied form in varied water depths is a testimony of its success till date and future will see deployment of these floating production system on much larger scale as future belongs now to deep and ultra deep waters where floating production system in conjunction with sub-sea production system offers a good cost-effective field development solution.

FLOATING PRODUCTION SYSTEM: ADVANTAGES

Looking from the perspective of small fields, marginal fields and the fields in deep and ultra-deep water, floating production system have a number of advantages over a fixed type platform based production system. The floating process facilities are selected over permanent production/drilling facilities in order to lower the financial risks involved. It is preferred in some cases when reserves were proved to be insufficient for a conventional development. The FPS gives minimum operating problems and is considered as standard development systems for marginal fields in shallow and calm waters. Normally, the floating production systems are available on lease. The use of a leased floating process facility minimized the initial capital cost, accelerates the start-up date and also enables maximum recovery of equipment in the event the reservoir performance did not meet the expectation. Floating Production Systems enables development of fields with little infrastructure for oil export and offers not only a low installation cost but also a low decommissioning cost at the end of the field life. They offer the advantage of being deployed readily elsewhere.

Further, Floating-type offshore structures have got a lot of cost advantage over fixed type platform structure. As we go for the development of oil and gas fields in deep and ultra deep waters having more than 1,000 m depth, feasibility of fixed platform structures often becomes negative and hence unacceptable owing to higher costs which often increases exponentially as we move deeper, mostly owing to the increase in the involved weights. Hence, floating type production systems are preferable in such water depths. Besides this, floating-type offshore structures are useful to produce oil and gas in marginal fields, that is, for a shortened production period. So one can plan exploitation of a number of small fields with a single FPS either simultaneously or one after another. FPS can also be designed, built, transported to the site, installed, and commissioned fairly rapidly; and removed, modified, and moved to other similar applications as needs change.

To summarize, Floating Production Systems has following advantages over fixed production system:

- Lower capital cost compared with a fixed platform
- Reduced time from discovery to production (i.e. early production)
- The FPS can be relocated and reused in another field
- Ability to operate in deep and ultra-deep water
- Can be used for reservoir testing in different locations
- Possible use in earthquake or ice-infested areas.

FLOATING PRODUCTION SYSTEM (FPS): TYPES

Floating Production Systems (FPS) are basically of the five types:

1.1 FPSO: Floating Production, Storage and Offloading Systems; Ship type vessel
1.2 FSO: Floating Storage and Offloading; Barge type vessel
1.3 Semi-submersible: A floating production facility
1.4 TLPs: Tension Leg Platforms: Conventional as well as Mini-TLP
1.5 SPAR: Deep draft caisson vessel.

Mobile Offshore Drilling Units (MODU) is the predecessor of floating production system, wherein it was used not only to develop the field but also to produce from it to start a limited production after development wells are drilled and completed.

All these five FPS differ on many accounts, one primary being the nature of buoyancy. The first three, viz. FPSO, FSO and FPS are "neutrally buoyant" (which allows six degree of freedom) type structures intended to cost-effectively produce and export oil and gas. Since these structures have appreciable motions, all wells are typically sub-sea completed and connected to the floating units with flexible risers (either of a composite material or rigid steel with flexible configuration i.e. compliant vertical access risers). Minimizing Deck-Payload and the overall size and displacement of units are the main considerations. The other two, viz. TLPs and SPARs, are 'Positively Buoyant" and they have limited motions and provide a suitable facility for surface-completed wells.

Before we discuss each type of floating production system in some detail, let's re-capitulate that any floating type offshore structure must posses the following:

- Appropriate work area, deck load capacity, and possible storage capacity
- Acceptable stability and station-keeping during harsh environmental actions
- Sufficient strength to resist harsh sea and environmental actions
- Durability to resist fatigue and corrosion actions
- Possible capabilities needed for both drilling and production
- Mobility as and when needed.

1.1 FLOATING PRODUCTION, STORAGE AND OFFLOADING VESSEL (FPSO)

A Floating Production, Storage and Offloading vessel (FPSO), is a ship based structure used in the offshore oil and gas industry and is designed to take all of the oil or gas produced from a nearby platform (s), or sub-sea production system, process it, and store it until the oil can be offloaded to waiting tankers, or sent through a pipeline. FPSO consists of a vessel or hull that normally stores the oil besides performing other functions and a host of topsides facilities put on its deck to process the oil or gas. The FPSO is kept in place at sea by a suitable mooring system. A FSO is a similar system, but without the possibility to do any processing of the oil or gas. An FPSO has the capability to carry out some form of oil separation process thus obviating the need for such facilities to be located on an oil platform. Oil is accumulated in the FPSO until there is sufficient amount to fill a transport tanker, at which point the transport tanker connects to the stern of the floating storage unit and offloads the oil.

FPSO can utilize varied mooring systems (any device used to hold secure an object by means of wire ropes, nylon ropes, chains). Some mooring systems used with FPSO are Internal Turret mooring, External Turret mooring, Yoke Turret mooring, and Spread mooring. Early FPSOs in shallow waters and in mild environment had spread mooring systems. But now a days, as more FPSOs were designed and constructed or converted (from an oil tanker) for deepwater and harsher environments, new more effective mooring systems were developed including external and internal turrets. Some turrets were also designed to be disconnect able so that FPSOs could be moved to a protective environment in the event of a hurricane or typhoon.

A Schematic diagram of FPSO is given in the next page.

1.1.1 FPSO Specification

FPSO is specified and represented in a particular way for a particular field and field-applications. Normally, any FPSO specifications have the following elements:

- Type of FPS
- Field of deployment
- Field conditions
- Water depth where it is deployed
- Size of FPSO in deadweight tonnage
- Storage capacity of the FPSO in barrels
- Uptime record in percentage
- Whether owned or on lease; if on lease, then period of lease along with the externsion period, if given
- Number of risers and umibilical
- Description of the swivel stacks, if there, specifying the number and sizes the lines of water injection, oil production, production testing, gas lift/export gas, fire water, power and control and the lines for hydraulic and utilities and for future options
- Maximum throughput from FPSO in barrels of oil per day
- Number of wells (producers, injectors- gas lift and water injection)
- Provisions of gas lift in MMSCFD; whether yes or no
- Provisions for water injection in bbls of water per day
- Date of first oil.

Traditionally for the FPSO construction, the hull is built in a shipyard and topside facilities in modularised form in a separate yard. The modules are then transported, lifted and placed on the top of the FPSO hull deck. These modules are therafter integrated into a comprehensive whole. Pre-commissioning and commissioning exercises follows thereafter. All this requires highest

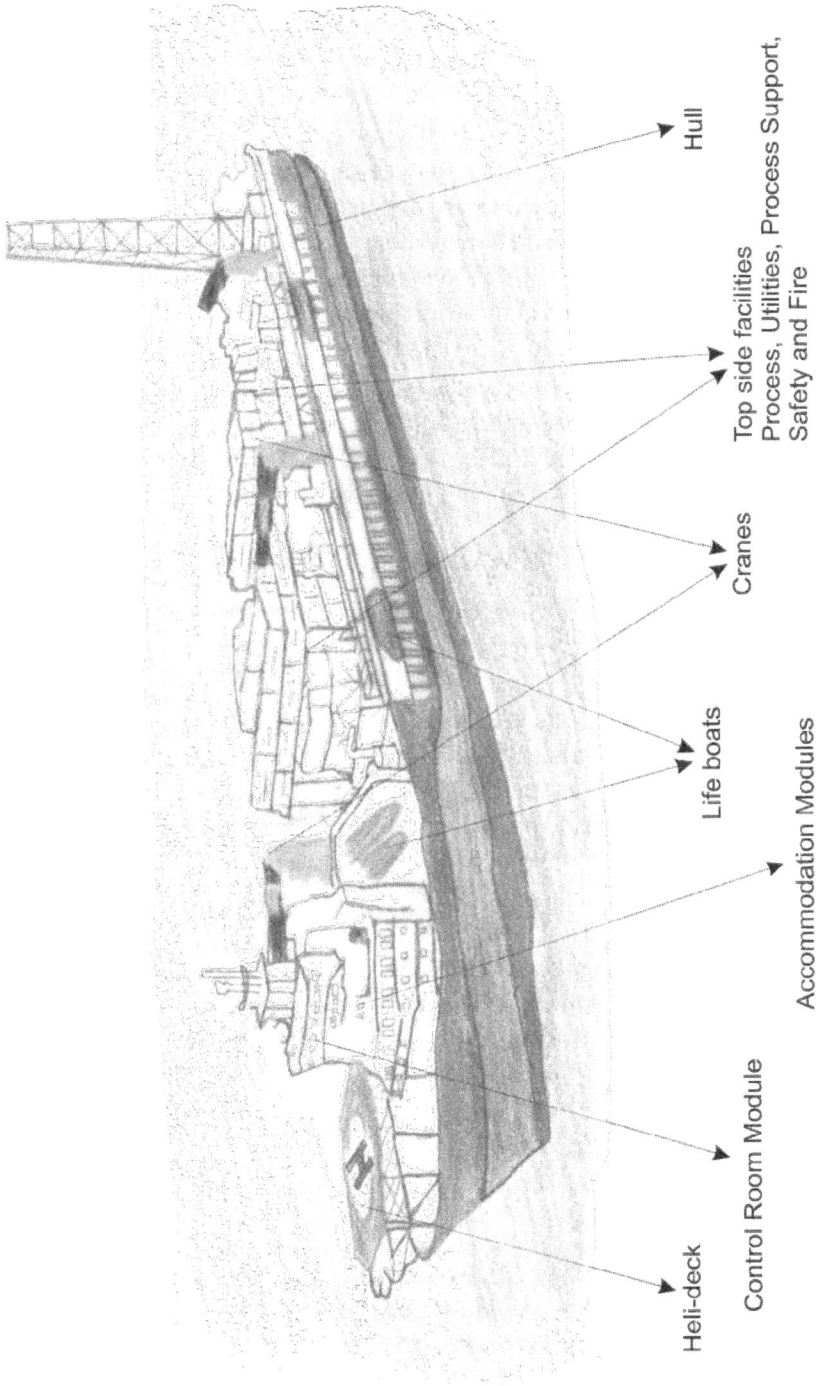

Hull

Top side facilities
Process, Utilities, Process Support,
Safety and Fire

Cranes

Life boats

Accommodation Modules

Heli-deck

Control Room Module

FPSO: Schematic Diagram

degree of project and general management besides adopting of a number of simultaneous operations to tackle time and cost dimensions.

1.1.2 FPSO: Advantages

FPSO has the same advantages as discussed above in general for floating production system. FPSOs are particularly effective in remote or deepwater locations where seabed pipelines are not cost effective. FPSOs eliminate the need to lay expensive long-distance pipelines from the oil well to an onshore terminal. They can also be used economically in smaller oil fields which can be exhausted in a few years and do not justify the expense of installing a fixed oil platform. Once the field is depleted, the FPSO can be moved to a new location. As old tankers can be used as FPSO, so the initial cost can be low. However, FPSO is for oil-field use only because FPSO doesnot posses much of the advantages for gas fields. Besides this, FPSO incurs potentially high cost for well workover and also the high turret mooring costs for certain areas.

1.2 FLOATING STORAGE AND OFFLOADING UNIT (FSO)

A Floating Storage and Offloading unit (FSO) is, as its name suggests, a floating storage device, usually for oil, without having any processing capabilities. These Barge-based floating production systems are suitable for marginal field developments in calm waters and are usually deployed at water depths of 30 ft to 150 ft. FSO are used to receive production from both satellite platform and free-standing caisson wells. Most FSOs are old single hull supertankers that have been converted. FSOs are commonly used in oil fields where it is not possible or efficient to lay a pipe-line to the shore. The production platform or sub-sea production system transfer the oil to the FSO where it will be stored until a shuttle tanker arrives and connects to the FSO to offload it. FSO is normally connected to the the manifold of the production facility (either a platform or sub-sea) by a flexible hose through six-point mooring system. A more advanced FSO with some processing capabilities is called FPSO. Barring processing system and facilities, FSO has all the features similar to FPSO and hence basic understanding of both as a ship-based system remains the same.

1.2.1 FSO Specification

Like FPSO, FSO is also specified and represented in a particular way for a particular field and field-applications. Any FSO specifications must have the following elements:

- Type of FPS
- Field of deployment

- Water depth where it is deployed
- Size of FSO in deadweight tonnage
- Storage capacity of the FSO in barrels
- Uptime record in percentage
- Whether owned or on lease; if on lease, then period of lease along with the externsion period, if given
- Number of risers and umibilical
- Description of the swivel stacks, if there, specifying the number and sizes of condensate lines. As FSO doesn't have processing capability, so no specifications required for oil, gas or water injection lines
- Maximum throughput from FSO in barrels of oil per day
- Number of wells as Nil
- Provisions of gas lift as Nil
- Provisions for water injection as Nil
- Date of first oil.

1.3 SEMI-SUBMERSIBLE

A Semi-submersible platform or rig is a mobile structure used either for drilling or for production of oil and gas in offshore environments in water depths ranging from 600 up to 35,000 feet. We can also understand semi-submersible or submersible as a watercraft that can put much of its bulk underwater. In other words, semi-submersible is a deck, supported well above the sea above the highest expected waves by submerged pontoons, with large column spreads providing floating stability. These units transits afloat on pontoons and requires "stability columns" to safely submerge to a bottom founded mode of operation. Their superstructures are supported by columns sitting on hulls or pontoons which are ballasted below the water surface. They provide excellent stability in rough and deep seas. Semi-submersible platforms have legs of sufficient buoyancy so as to cause the structure to float. Semi-submersible also has the weight sufficient enough to keep the structure in upright position.

Semi-submersible rigs can be moved from place to place and can be ballasted up or down by altering the amount of flooding in buoyancy tanks. They can be towed into position by a tugboat and anchored, or moved by and kept in position by their own azipod propellers with dynamic positioning. Station-keeping of semi-submersibles is usually achieved by chain- or wire-mooring systems. Normally semi-submersibles are spread-moored. Where moorings are not practical, dynamic positioning systems with computer-controlled thrusters that respond to vessel displacements or accelerations are used. As the semi-submersible has a relatively small area above the water surface, the semi-submersible is less affected by the waves than a normal ship.

Platform Rig

Cranes

Wells/Well-bay

Living Quarter

Life boats

Heli-deck

Columns

Flexible risers, controls

Pontoons

Semi-submersible: Schematic

Deck loaded with process, utilities, safety and fire fighting facilities

Hull Type Structures Comprising of Columns + Pontoons + Bracing systems

Officially known as "column stabilised units", "Semi-submersible" (and also "submersible") has the "stability columns" that primarily provide floation stability. Semi-submersible are "column stabilised" meaning that the center of gravity is above the center of buoyancy, and the stability is determined by the restoring moment of the column. This contrast with the SPAR platform, which achieves stability by placing the center of gravity below the center of bouyancy and the TLP, whose stability is derived from the tendons.

One important thing to understand is that virtually all semi-submersible have at least two floation states: the one "afloat on the columns" when semi-submerged and the other one "afloat on the pontoons" when not semi-submerged. Further to understand is that though columns function structurally, providing structural strength is not the main functions of the column. Its main function is to provide floation and stability. Size, submergence, proportion and spacing of all the columns are major factors in the hydrodynamic performance of semi-submersibles.

Semi-submersible has normally been used mainly for drilling purposes, but since the early 1980s, these have also been used as production platforms without having any oil storage capacity. Besides these two main uses, Semi-submersible can also be used for a variety of functions like heavy lift, accomodations, as operational support (surface, subsea) and even for space launch. Important point here is that a function needs to be clearly spelt out beforehand whenever we are going for the design and configuration of semi-submersible. Further, semi-submersible essentially performs two functions. The first one is that it support a payload above the highest waves in a stablised way and understanding of this function helps in configuring the number, size and spacing of the stability column and the height of the deck. The second function is that semi-submersible provides minimum response to wave and understanding of this function helps in configuring the size, shape and submergence of pontoons relative to column waterplane area and the spacings of the pontoons and columns.

1.3.1 Semi-Submersible: Components

Any Semi-submersible has three basic components:
- Deck
- Multiple stability columns and Pontoons
- Space frame bracing

Deck

The deck provides the working surface for most of the semi-submersible's functions. Structurally, deck has a function to transfer the weight of the deck and its loadings to the columns (and bracings). Besides this, deck is also a part of the overall global strenghth system providing structural connections between all the columns. The pontoons and columns are generally arranged and connected in a way that can provide considerable global strenghth. To ensure this global strenghth, deck is arranged likewise and connected. Where this arrangement doesnot provide sufficient global strength, a space frame bracing system is employed. Bracing system, however, is avoided for the reasons that they are expensive to build, are vulnerable to fatigue and are a costly maintenance item in regard to inspection and repairs.

Deck has evolved over the time. Initially it was a single level structure with individual deckhouses arranged with no coherent interrelated structural functions. Later on, it has devloped into a "Hull-type superstructure" with integral connections to the columns tops. This hull-type integrally connected desk is superior in strenghth, has considerable usable interior space, and valuable floation in damaged stability. Further, if built with rest of the hull in a modern shipyard, a hull-type deck is lighter, less costly, and of superior strenghth than other alternatives. A disdavantage with hull-type is the necessity for mechanical ventilation and inability to get it built fully by a single builder. Another kind of deck is the "Truss-Deck". It is preferred in some production applications that favour open, natural ventilations as well as historical design and fabrication practices. Wherever there is separate fabrication, outfitting, and joining of the deck (spilit construction), the truss-type deck is preferred because most fabricators of production decks are not equipped to build plated-structures. So choice of the deck type is of considerable importance in configuring a semi-submersible in so far as it determines whether a split or an integrated construction will be preferred.

Multiple Stability Columns and Pontoons

Normally, Semi-submersible structures have two submerged horizontal tubes called pontoons, which provide the main buoyancy for the platform and also act as a type of catamaran hull when in transit to or from a site at low draft. A ring pontoon is also sometimes used solely for one fixed location. Typically, four to eight vertical surface-piercing columns are connected to these pontoons. The platform deck is located at the top of the columns.

The numbers and arrangements of pontoons and column are important consideration in the configurational variants of semi. There can be as few as three to as many as a dozen or more column. Likewise, there can be a simple two parallel pontoon arrangement, or upto six pontoon arrangement, and even a grillage of orthogonally intersecting pontoons. Besides this, there can be independent footing pontoons, one for each stability column. Practically, only the 4-, 6-, and 8-column configuration and only the twin pontoon and closed array pontoon arrangements are used.

Space Frame Bracings

Bracing system plays an important role in the selection of pontoons and arrangements type. A closed array or 'ring' pontoon arrangements is not good for towing mobility, but it is preferred for a permanently sited system owing to its superior strength and excellent potential for braceless system. In this cae, transverse braces are not required and with well designed column to pontoon connections as well as special connections at the deck, the system can handle the racking loads.

We have a number of different types of bracing configuration. Transverse bracing which is low on columns and which is used to resist sqeeze/pry forces have transverse diagonal bracing. Diagonal bracing is both to support the deck weight and together with horizontal transverse provide strength. Then we have a system of horizontal diagonals which are provided for racking strenghth against quartering seas. Longitudinal bracings are not used where continuous, strong longitudinal pontoons are employed. A well designed and well connected deck structure can eliminate the need for most bracing. Similarly, a closed array pontoon can also eliminate the need for bracing.

1.3.2 Semi-Submersible: Advantages and Disadvantages

The advantages of semi-submersibles are that Semi-submersibles can achieve good (small) motion response and, therefore, can be more easily positioned over a well template for drilling. Further, Semi-submersibles allow for a large number of flexible risers because no weathervaning system is involved. However, Semi-submersible has some disadvantages too. Though semi-submersibles have low initial costs, it incurs potentially high cost for well work-over. Further, in semi-submersible case, pipeline infrastructure or other means is required to export produced oil as most of the semi-submersible doesn't have oil storage. Another disadvantage is that only a limited number of (rigid) risers can be supported because of the bulk of the tensioning systems required. Also considering that most semi-submersible production systems are converted from drilling rigs, the topsides weight capacity of a converted semi-submersible is usually limited. Further, building schedules for semi-submersibles are usually longer than those for other ship-shaped offshore structures like FPSOs or FSOs. Semi-submersible poses significant design problems too. The main problem in semi-submersible design is to adopt the right configuration for the specific functions required and the construction programme.

1.4 TENSION LEG PLATFORMS (TLP)

Tension leg platforms are the vertically moored floating structure normally used as drilling and production platforms at water depths greater than 1000 ft. TLP is a fixed-draft constant buoyancy system essentially having the function of supporting a payload above the highest waves. TLP once installed, doesnot rely on floatation stability. The platform is permanently moored by means of tethers or tendons grouped at each of the structure's corners. A group of tethers is called a tension leg. TLP structures are attached to the bottom with tendons held in tension, that's why called Tension Leg Platform. TLP consists of columns, pontoons and a mooring system which consists of vertical tendons (called tethers) and which is used to restrain the heave motion.

In comparison to a semi-submersible, which is a true, free floating structure restrained with compliant spread mooring and/or dynamic positioning, a Tension Leg Platform is kept in place through lateral forces developed by the tendons when the TLP is moved off from center. The lateral force is dependent upon the tendon tensions. Hence, a major portion of the buoyancy of any TLP is devoted for the development of tendon tension. Besides this, while dynamic mooring loads of other floating structures are largely mitigated by platform inertia, the mooring loads of TLPs are directly linked to first order wave loads on the structure.

1.4.1 TLPs: Components

TLP has the following components:

- Pontoons
- Stability Columns
- Deck
- A mooring system comprises of Tendons (Tethers)
- Hull

Normally, for a TLP, decks have always been long span four-point corner support open trusses. As such the decks are multilevel structures. The single column and close-clustered multi-column decks resemble like a fixed platform with a four point support as if a 4-pile jacket support it. All decks of TLPs are built from the hull and joined later either at a dockside, offshore, or in a separate sheltered location.

A TLP may have up to six vertical surface-piercing columns with a complete ring of pontoons and a number of vertical tethers. TLPs have relatively large motions of surge, sway, and yaw whereas it has usually small heave, roll, and pitch motions (motion in comparison to significant wave periods). And these

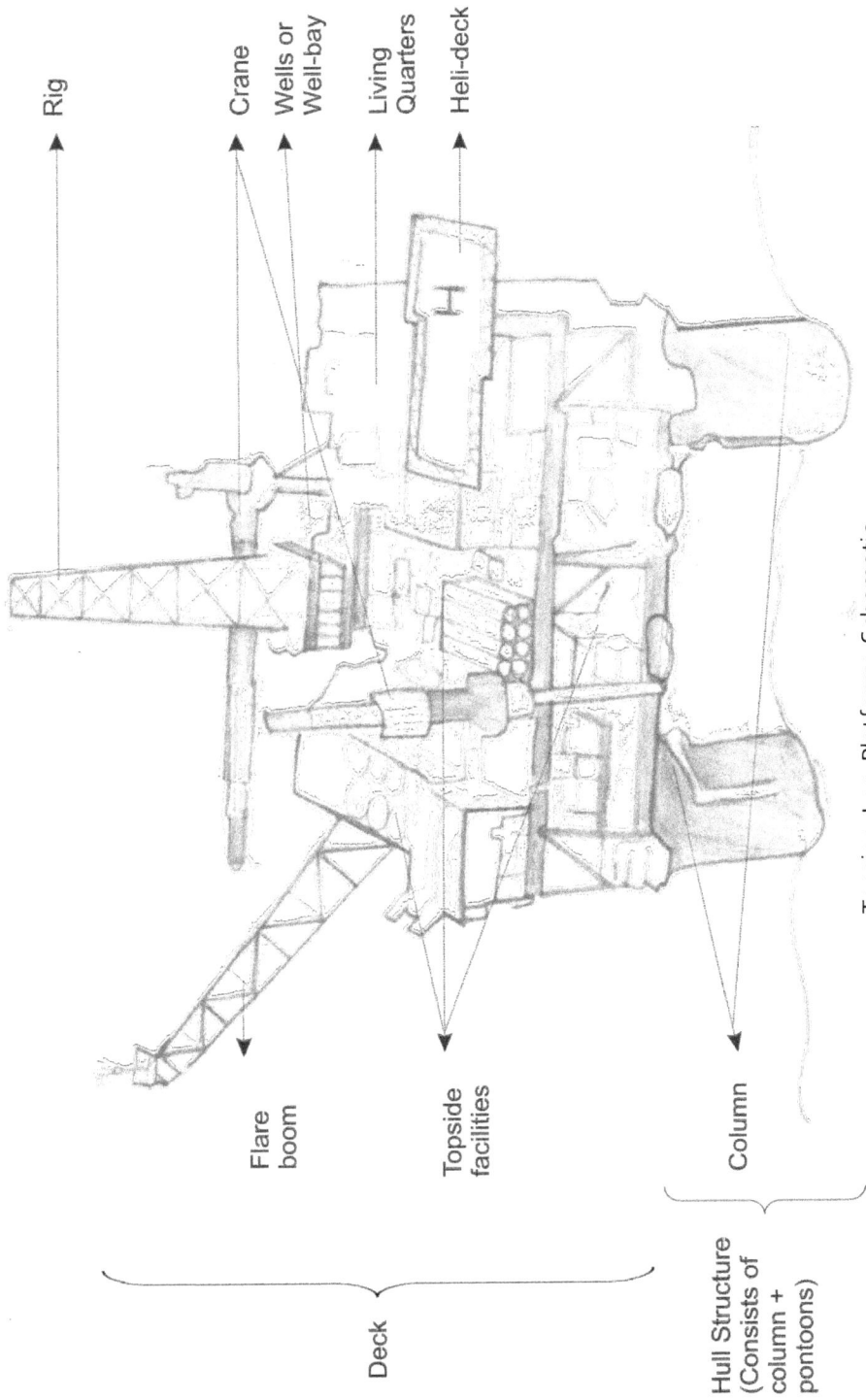

Rig

Crane

Wells or
Well-bay

Living
Quarters

Heli-deck

Flare
boom

Topside
facilities

Column

Deck

Hull Structure
(Consists of
column +
pontoons)

Tension Leg Platform: Schematic

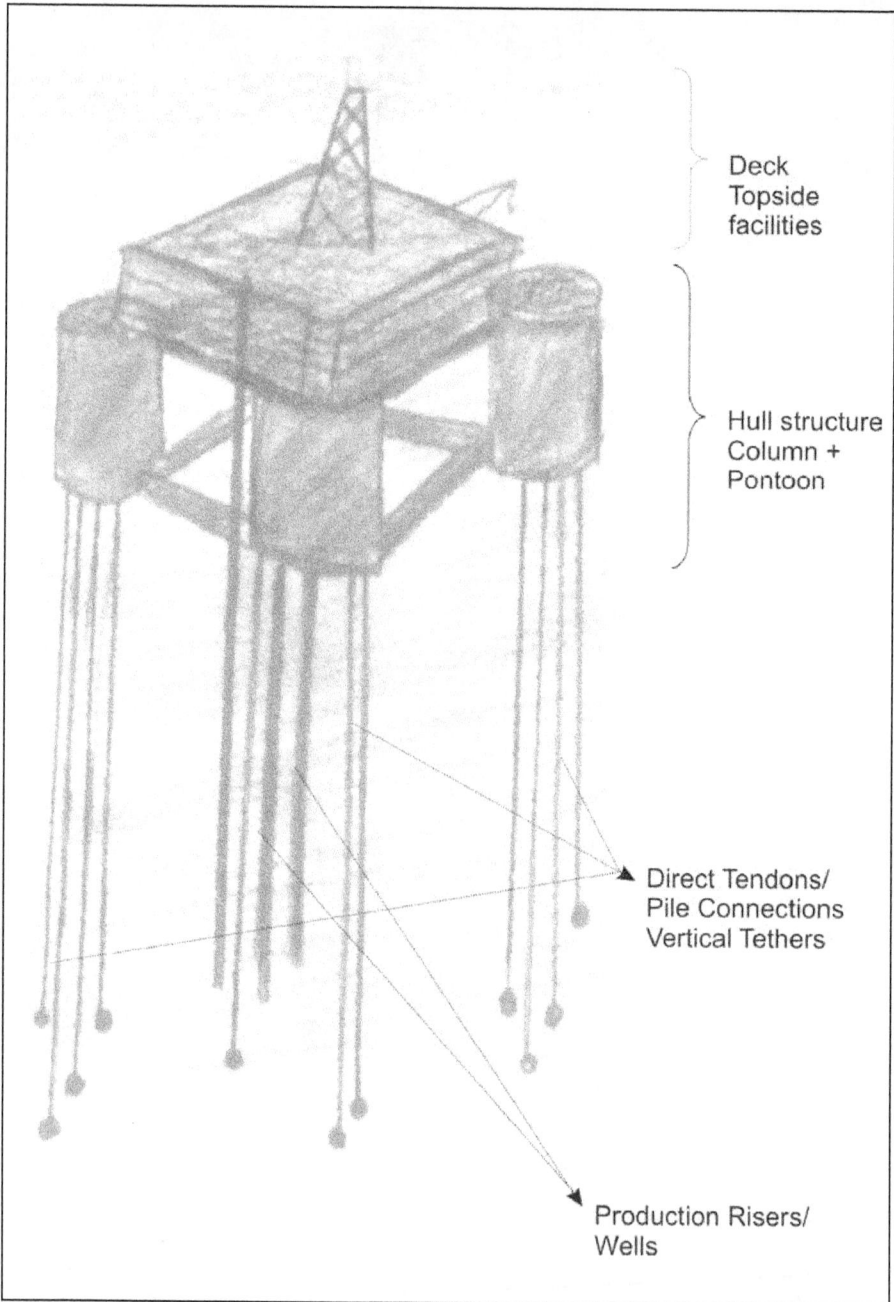

Deck
Topside
facilities

Hull structure
Column +
Pontoon

Direct Tendons/
Pile Connections
Vertical Tethers

Production Risers/
Wells

Tension Leg Platform; Schematic

motions are quite affected by the design and configuration of tethers. In TLPs, columns are normally the principal source of boyancy with the pontoons functioning more as structure. The pontoons and the columns bear a size relationship in regard to hydrodynamic force. Further, in operational state, TLPs employ very little ballast. Ballast is provided to even loading between tendons and also to offset unused payload capacity.

Tethers are one very important component of any TLPs. Tethers are designed in such a way that they have relatively high axial stiffness (low elasticity), thereby eliminating virtually all the vertical motions of the platform. This allows the platform to have the production wellheads (dry x-mas tress) on deck (connected directly to the subsea wells by rigid risers) instead of having them on the seafloor. This facilitates for a cheaper well completion and gives better control over the production from the oil or gas reservoir. Hull is another very important component that provides buoyancy both for the support of weight and for providing tendon tensions. Hull should be tall enough to give the deck wave clearance in all modes of operation. Both tendon tension and payloads have substantial influences on hull size.

1.4.2 TLPs: Advantages and Disadvantages

TLPs are exclusively used as permanently sited platforms. This permanent sitting reduces the maintenance cost. TLPs have comparatively low operating costs. As TLPs have a minimal range of functions, they have very little configurational dependence upon function. Further, TLPs doesnot have an inherent ability in itself to provide a petroleum storage option. Further, TLPs doesnot have heave motion because as far as heave is concerned, TLP is "fixed". TLPs are very stable with almost no heave, pitch or roll motion. TLPs provide easy direct well access through dry christmas tree. TLPs can have both the "top tensioned risers" and "steel catenory risers" with sub-sea tie-backs. Further, TLPs can be scaled down to small fields.

TLPs are laterally compliant and hence, surges, sways and yaws. This means that TLPs are highly compliant to lateral forces and at the same time, are highly resistant to vertical forces. One very important aspect of TLPs is that it undergoes set-down with offset. Set-down is a corrosponding downward motion coupled geometrically so as to do offset.

The disadvantages with TLPs are that, TLPs cannot be moved from location to location. TLP has a water depth limitation too (2000 meters) with steel tendons. Also, the sizes of TLPs present certain operational logistics problems too. Further, as TLPs are sensitive to payloads because of the tensioning effect of tethers, they are not usually used as storage units. Because of this, TLPs normally requires a pipeline infrastructure or FSOs plus a shuttle tanker offloading system to export the produced oil.

1.5 SPAR

Spar is an anchored buoy-like vessel that floats vertically and has on its top the production, treating and storage modules. Spar is nothing but a buoyant structure that has got a shape like a spar (a single, large-diameter cylinder) with a functional deck mounted on the top of it. A spar usually has a vertical circular cylinder with a very large diameter, say up to 80–150 ft and height up to 700 ft. This diameter-height proportion contributes significantly in the reduction of heave motion of the unit by virtue of the large draught. The Spar along with TLPs is the only floating production platforms that possess small enough heave and pitch motion to allow the risers to be safely and economically supported by the floater. A production spar may or may not have oil storage and related wells at surface or sub-sea.

Spars are usually moored in position by traditional spread mooring with drag embedded anchor so as to allow motion of all six degrees of freedom. The mooring system of a Spar can also be by a chain-wire-chain taut catenary system. A taut mooring system is defined as the one in which the anchor loads have an uplift component for all load conditions i.e the anchor chain or wire never lies on the seabed. The taut system saves a considerable length of wire and chain needed for a conventional catenary mooring. With taut mooring, the spar motion are small enough even in the 100 year hurricane. Alternatively, a tether-mooring system that makes it into a kind of tension leg platform with a single cylinder may be used.

Spars have been used for decades as marker bouys and for gathering oceanographic data. In the beginning, spars were used as storage units, but spars are also now used for production. Till date, there are three types of production Spar that have been built: The Classic Spar, The Truss Spar and the Cell Spar. In cell spar, a cell i.e a bundle of tubes, are extended down to the ballast section and can include plates. This cell spar is suitable for smaller economically challenged fields. In Truss spar, there are several horizontal plates and the lower part of it possesses much lighter truss structure. Further, this truss spar has the reduced current loading on moorings.

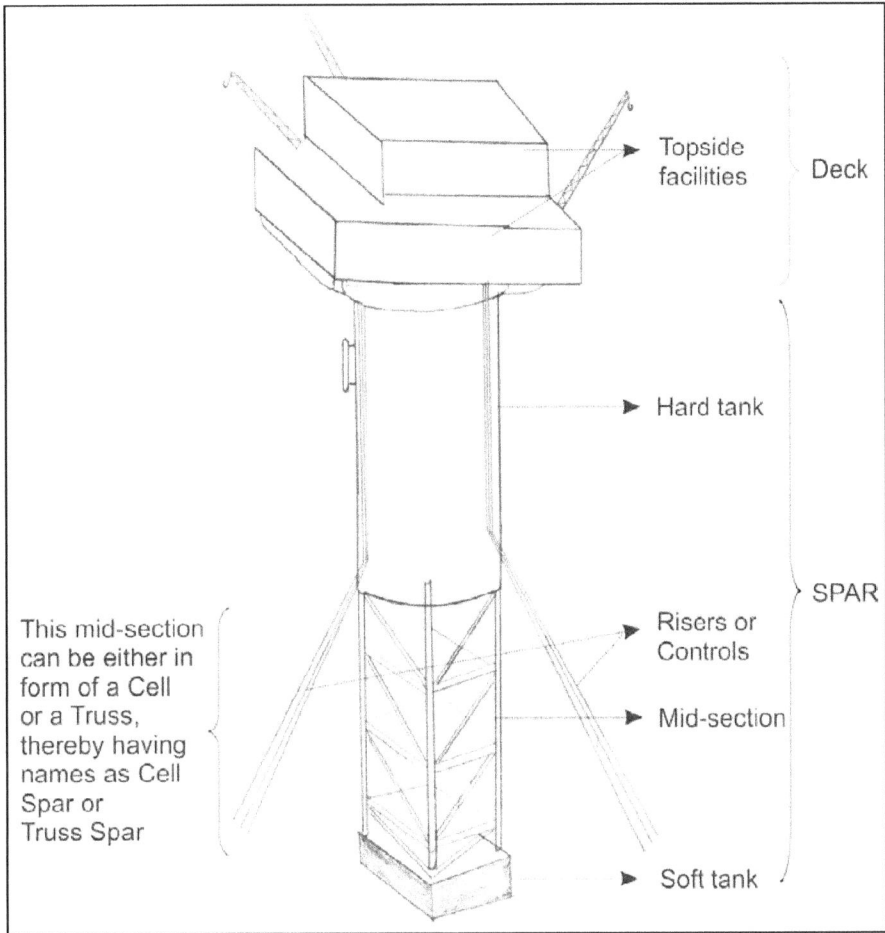

SPAR: Schematic

1.5.1 SPAR: Components

The basic components of Classic and Truss Spar are as follows:

- Deck
- Hard Tank
- Midesection (steel shell or truss structure)
- Soft Tank

Deck

Topside deck of the spar is typically a multi-level structure in order to minimise the cantilevel requirement. The deck weight is supported on columns (for upto 18,000 tons, four columns; additional column for heavier decks) which join the hard tank at the intersection of a radial bulkhead with the outer shell.

Hard Tank

The hard tank provides the buoyancy to support the deck, hull, ballast, and vertical tendons (except the risers). The term 'Hard Tank" means that its components are designed to withstand the full hydrostatic pressure. There are typically five to six tank levels between the spar deck and the bottom of the hard tank, each level seperated by a watertight deck. Each level is further divided into four compartments by radial bulkheads emanating from the corner of the centerwell. The tank level at the waterline includes additional cofferdom tanks to reduce the flooded volume in the event of penetration of the outer hull from a ship collision. Thus there are upto 28 separate compartments in the hard tank. Typically, only the bottom level is used for variable ballast, the levels being void spaces.

Midsection

Midsection extends below the hard tank to give the spar its deep draft. In the early "classic" spars the midsection was simply an extension of the outer shell of the hard tanks. There was no internal structure, except as required to provide support for the span of risers in the middlesection. The scantlings for the midesction were determined by construction loads and bending moments during upending. Later spars replaced the midsection with a space frame truss structure. This truss spar arrangement resulted in a lower weight, less expensive hull structure. Also the truss has less drag and reduces overall mooring loads in high current enviornments.

Soft Tank

The soft tank at the bottom of the spar is designed to provide floation during the installation stages when the spar is floating horizontally. It also provides the compartments for the placement of fixed ballast once the spar is upended. The soft tank has a centerwall and a keel guide which centralises the risers at that point.

1.5.2 SPAR: Advantage and Disdavantages

As far as the advantages of spars are concerned, spars allow direct well access through "top tensioned risers", supported by air cans or tensioners. Spars have small heave motions. Spars allow catenary moorings thereby requiring no tendons. Spars are insensitive to topsides weight. Further, spar has lower cost than TLP for very deepwaters. The disadvantage with spar is that it doesnot have any oil storage system. Further, in spar, large hull size might cause fatigue of many components like air cans, risers, moorings etc.

Floating Concept Selection Flow Diagram

Main Drives

Export pipeline or shuttle
Dry or subsea trees

Secondary Drives

Storage volume
Environment
Field life

KEY

◇ Yes/No Decision box
▭ Information box
▢ Concept description box

Start

Is pipeline infrastructure available for export?

— NO —

Are dry trees required?

— NO —

Are dry trees required?

Yes → Use a TLP or SPAR with an FPSO or a SPAR with storage

NO → Is it a benign environment

Yes → Use a spread moored new built barge or converted tanker

NO → Use a turret moored new built FPSO or a converted tanker

Is field life >20 years?

Yes → Is required storage >100,000 bbls?

NO → Consider a semisub

Yes → Is field life >15 years?

Consider a semi-sub FPV with limited storage

Wellhead semi-sub with FPSO

New build spread moored production barge or new tanker convention (intercept)

Use an old converted tanker

Use a leased spread moored FPSO

Use a new-built turret moored FPSO or new converted tanker (intercepted)

Use a leased turret moored FPSO

Is field life >15 years?

Is required storage >100,000 bbls?

Yes → Use a wellhead TLP or SPAR with a FPSO

NO → Use a Wellhead SPAR with a FPSO

Is water depth >1200 m?

Consider a SPAR with storage

Yes → Use a Semisub or Barge

Are dry trees required?

Yes → Use a TLP or SPAR

Is water depth >1200 m?

Yes → Use a SPAR

NO → Use a TLP or SPAR

Is it benign environment?

Yes → Use a spread moored Barge with no storage

NO → Use a semi-sub FPV

Use a Monohull and/or semisub

Use a spread moored new built barge or converted tanker

Mooring Systems

Mooring is an important aspect of any floating production system. Ships or Vessels in the sea have to be kept in its place to enable them either to produce oil and gas and/or to take transfers and storage of oil and gas on board. This permanent or temporary station keeping at the sea is called "to moor a ship or vessel". A vessel is said to be moored when it is fastened or hold secure to a fixed object such as a pier or quay, or to a floating object such as an anchor buoy by means of cable, anchors or lines. Every sea bound vessel, thus, has a mooring system, the extent or design of which depends upon a host of factors like client requirement, water depths, sea conditions, environmental conditions, process requirements, load requirements, etc.

The one very important consideration while designing a mooring system of a vessel is to consider whether the vessels are allowed to "weathervane" or not. Weathervane means the "Turning of a vessel at anchor from one direction to another under the influence of winds or currents". It means that the ship can rotate in the horizontal plane (yaw) into the direction where environmental loading due to wind, waves and currents is minimal.

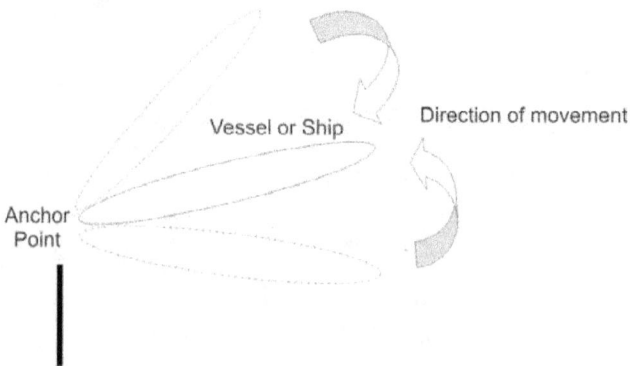

Vessel or Ship   Direction of movement

Anchor
Point

"Weathervane" concept

Mooring is often accomplished using thick ropes called mooring lines or hawsers. The lines are fixed to deck fittings on the vessel at one end, and the

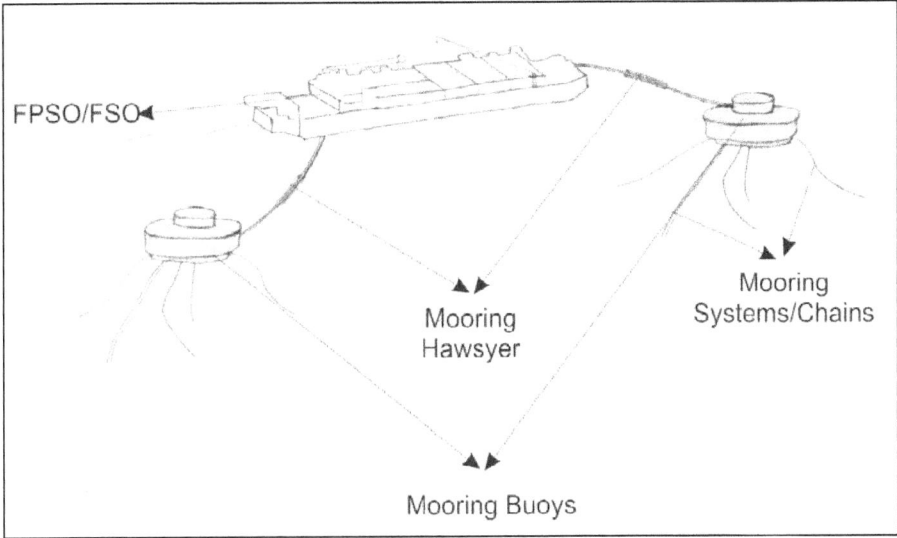

Mooring System

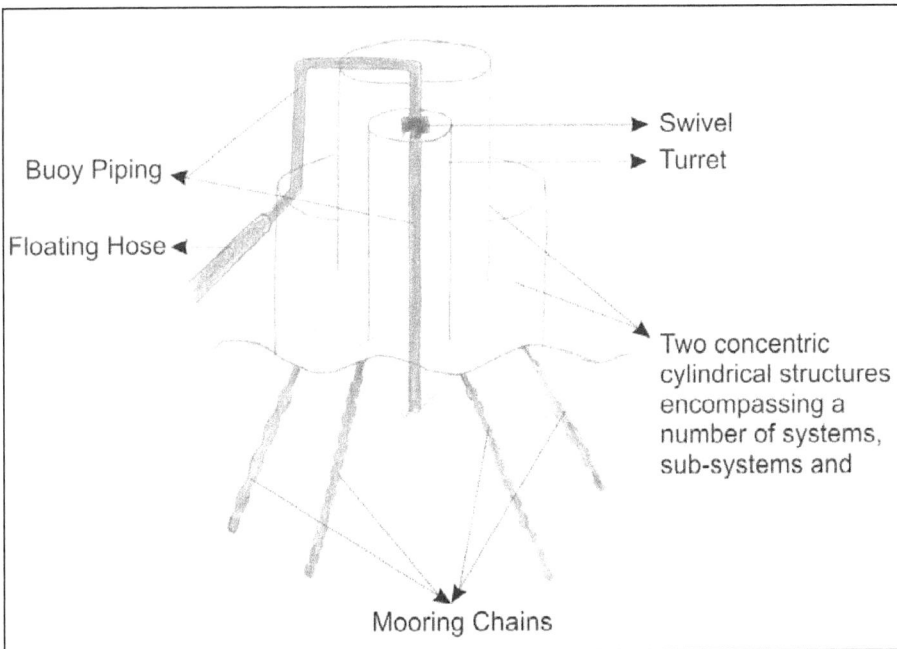

Mooring Components

other end is fixed to the bouy. Complete hawser arrangement includes nylon rope, shackles, chafe chain, load pins and marker buoys and are usually purchased as full assembly to ensure better integrity. Mooring, however, is also achieved by deploying a permanent anchor. In mooring schemes, the bow and aft size of the tanker or ship has different specific purposes as mentioned below:

Position	Name	Purpose
1.	Bow line	Prevent backwards movement
2.	Forward Breast line	Keep close to pier
3.	Forward spring line	Prevent from advancing
4.	After Spring line	Prevent from moving back
5.	After Breast line	Keep close to pier
6.	Stern line	Prevent forwards movement

Aft side of the ship   Bow side of the ship

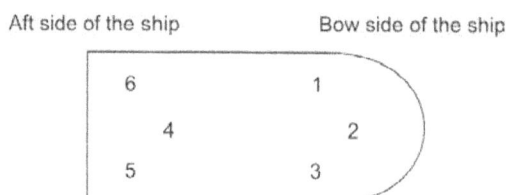

2.1 MOORING TYPES

Mooring systems can broadly be categorized into four types:

2.1.1 Single Buoy Mooring System
2.1.2 Turret Mooring System
2.1.3 Spread Mooring System
2.1.4 Tower Mooring System

On a general note, Single point mooring tend to be used more frequently for ship shaped vessels, whereas the spread mooring system is generally more used for semi submersibles and spars. However, we will discuss specific use patterns of these different mooring systems along with their mooring arrangements and their sub-types in the following paragraphs. These mooring systems utilize a number of mooring accessories to work as a comprehensive whole with good integrity. These accessories are discussed in brief later in this chapter.

2.1.1 Single Buoy Mooring (SBM) or Single Point Mooring (SPM)

Single Buoy Mooring (SBM), also known as Single Point Mooring (SPM), comprises of a round buoy (a very large diameter round float) fixed to the seabed by chains or anchors. The buoy body provides the necessary buoyancy and stability to the entire unit. These buoys acts as moorings points for the tankers and also essentially performs the task of fluid transfer between the offshore producing facilities and the moored tanker.

Arrangement wise, this mooring system has a top portion (buoy body) comprising of a turntable, bearings, product swivel, buoy piping around a turret and a host of equipment pertaining to product transfer, lifting and handling besides having the space for boat landing, fendering, power provisions and navigational aids. The buoy body provides the necessary buoyancy and stability to the entire unit. This top portion is held in position through mooring and anchoring components comprising of anchors or piles, anchor chain and chain-stoppers. The tanker moors onto the buoys by means of a hawser arrangement system (nylon rope, shackles, chafe chain, load pins and marker buoys) and rotates by any force of currents or winds into the path of least resistance. The product transfer system comprises of flexible sub-sea hoses (called Risers), floating hose strings, product swivel, valves and piping. Riser configuration is important to understand here as these tends to vary depending upon water depth, sea conditions, buoy motions etc. Product swivel provides one or multiple independent paths for fluid, gases, electrical signal or powers etc. The oil swivel (roller bearing) at the center of turntable allows the moored tanker to weathervane freely around the buoy (say 360 degree rotation) without impeding the oil flow.

Depending upon whether we are using the mooring hawser or the rigid arm to connect the tanker to the buoy, we have two types of single buoy mooring systems: Catenary Anchor Leg Mooring (CALM) buoy that uses mooring hawser and Catenary Anchor Leg Rigid Arm Mooring (CALRAM) buoy, the basic principle remaining the same in both the types. Another buoy mooring system is the Conventional Buoy mooring (CBM) which basically is a spread mooring system consisting of multiple buoys. However, this CBM does have the limitation of non-weathervaning, thereby creating substantial limitations on operations and deployment.

The SPM or SBM is most widely used because of its relatively low operational costs, flexibility and reliability and also because of its ability to berth tanker of any size including the ULCC (Ultra large crude carriers) and VLCC (Very large crude carriers). This mooring system can be chosen for short term mooring, permanent mooring as well as permanent mooring with easy disconnect capability so as to evacuate the facility in case of severe weather conditions. Deployment of this kind of mooring system requires comprehensive pre-deployment studies to understand behavior of the buoy in various the sea-conditions (winds, wave, and current), water-depths, tanker

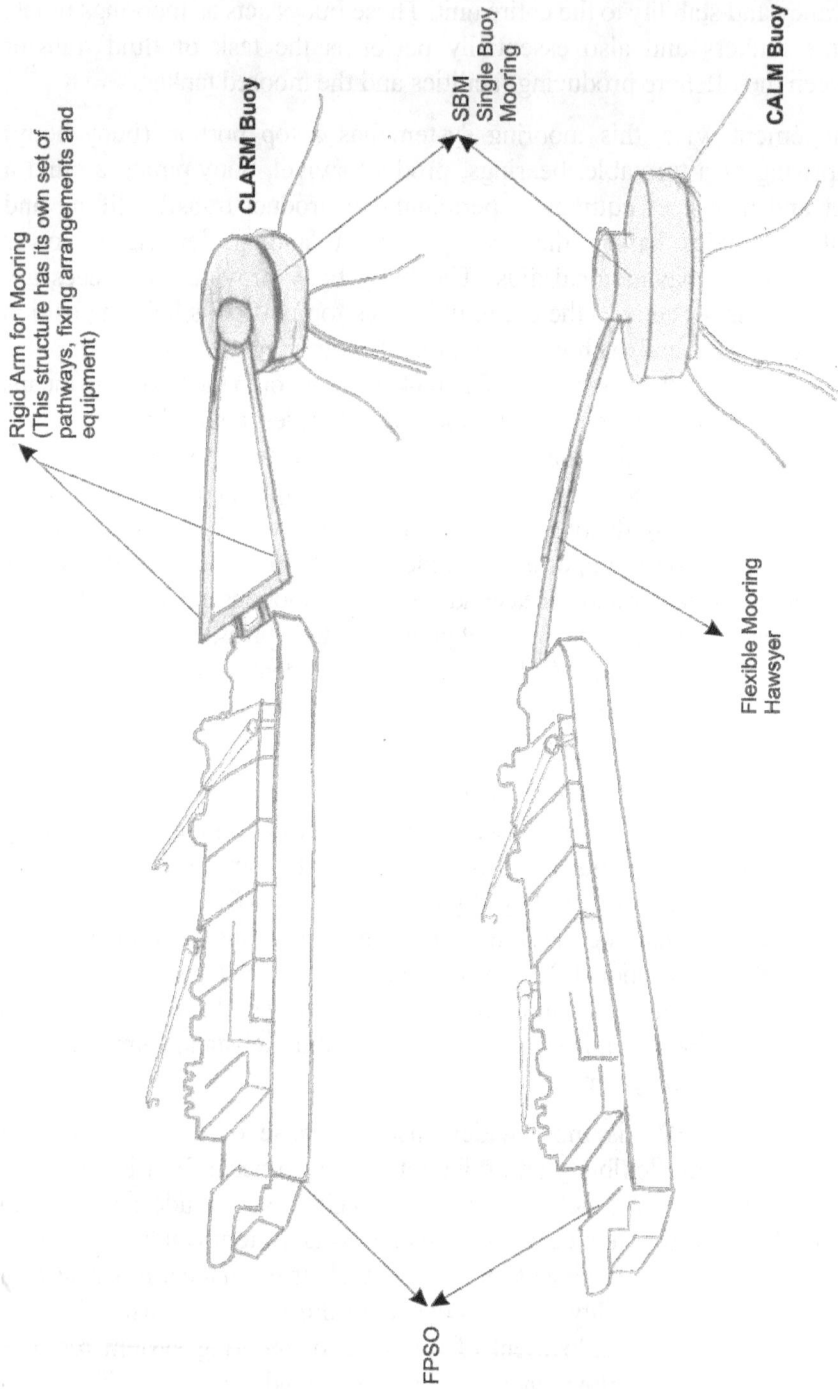

CALM Buoy and CLARM Buoy

sizes etc so as to determine the "optimum" w.r.t. the arrangement and sizes of mooring legs and its various components. Soil conditions also need to be studied to ascertain the optimum anchorage points.

2.1.2 Turret Mooring System

Turret mooring systems are the single point mooring systems that permit the vessel to freely "weathervane" 360 degrees thereby allowing normal operations in moderate to extreme sea conditions. Turret mooring system has got its name because of the use of a turret structure (turret means a small tower or tower-shaped projection having great height in proportion to its diameter and having an ability to revolve or rotate). In this type of mooring system, lines are connected to the turret which via bearings allows the vessel to rotate around the anchor legs.

The turret mooring system comprises of a turret structure (external or internal), swivels (fluid, gas, electrical power and control swivel), chaintable (either above or below the waterline) and limited number of risers (say up to 100) and associated umibilicals. However, this mooring system doesn't have hawsers and floating hoses.

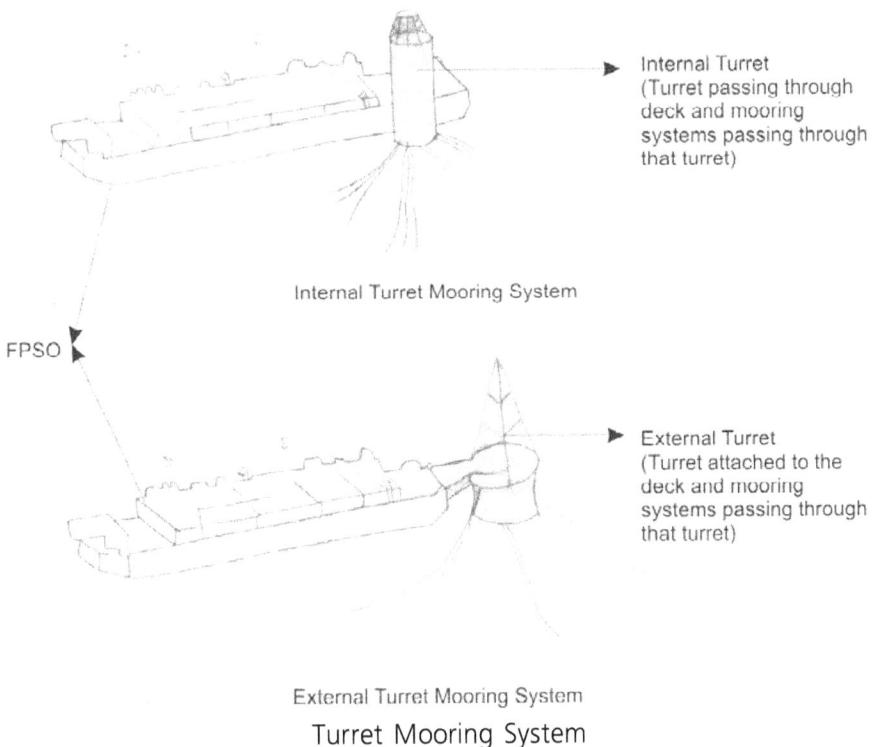

Internal Turret
(Turret passing through deck and mooring systems passing through that turret)

Internal Turret Mooring System

FPSO

External Turret
(Turret attached to the deck and mooring systems passing through that turret)

External Turret Mooring System

Turret Mooring System

Depending upon the position of turret mounting, this type of mooring system can either be "External turret mooring" or "Internal turret mooring". Apart from turret mounting, the configuration of this mooring system further depends upon a given application depending upon process requirements, vessel size, and sea and enviornmental conditions and hence it varies from application to application. In external turret mooring, turret is fixed to bow or stern of the ship whereas in the internal mooring system, the turret is placed within the hull. Out of these two, external turret systems are preferred because of its being less expensive, its having comparatively lesser design concerns, its ability to get delivered in a shorter period of time, no-dry-docking requirement, maximisation of the hull storage capacity and because of its wider application options in moderate to exreme sea conditions. For locations with severe cyclonic weather, harsh sea conditions and icebergs, disconnectable internal mooring systems are deployed, wherein vessels gets connected or disconnected easily in dangerous sea conditions like typhoons, hurricanes, icebergs..etc so as to avoid anticipated damage and to act safe.

2.1.3 Spread Mooring System

The spread mooring system is the simplest way of mooring an FPSO by deploying the traditional shipboard mooring equipment in such a way so as to keep FPSO or tankers at a fixed position and essentially with a constant heading (meaning that the heading of the system is independent from the direction of wave and winds). In this mooring system, sub-sea hoses are directly connected from the seabed to the FPSO or Tanker's manifold.

This mooring system utilizes a set of anchor legs and mooring lines, normally arranged in a symmetrical pattern, attached to the bow and stern of the vessel in a very simple way so as not allow the floater to "weathervane", hence, ensuring the fixed or constant heading for the vessel on site. However, for a temporary mooring and in moderate environmental conditions, conventional buoy or multiple buoys are used along with the anchors arranged in such a way so as to restrict the weathervaning. This absence of weathervaning capability does, however, increases the environmental load and thereby an increasing mooring lines or tension lines thereby making any alteration exercise to the hull of the vessel for attaching the mooring lines a cumbersome operation.

This mooring system is the best option on a site where the prevailing severe weather is highly directional and also for the applications requiring long service life, in any water depth, and on any size of vessel. This fixed heading capability of spread mooring system removes the requirement of turret structure and bearing and requirement of swiveling components and that of the requirement of hydraulic power, electrical power and control transfer. However, deployment of this kind of mooring system requires comprehensive pre-deployment studies to understand the sea-conditions, water-depths, soil conditions etc so as to determine the "optimum" w.r.t. the arrangement and sizes of anchor legs and its various components and also w.r.t. the length and strength of the sub-sea hoses or strings. In general, the anchor size for this mooring system is larger than those of turret, tower or buoy mooring systems.

2.1.4 Tower Mooring System

Tower Mooring System (TMS) is a single point mooring system in the form of a rigid tower-shaped structure. This essentially has two components, structurally:

- The tower has a lower end pivotally mounted to the seafloor with the help of a group of long chains anchored at the location spaced around the tower on the sea-bed. The positioning of chains is such that it helps in effective maneuvering of tower w.r.t. the tilting in any direction. Normally, most of the chain length lies on the sea-floor in the vertical tower condition.
- The upper end of the tower inclusive of a turntable lying above the sea-surface and connected through a mooring hawser (or through a wishbone) to a ship. Turntable helps the moored vessel to weathervane freely about the tower.

Besides this, the upper end of the tower has enough deck space for manifolding, pigging and for other auxiliary equipment required for transfers of product to the connected ship. This mooring system also has some device, say a yoke arm, which is connected to the turntable through pitch and roll joint that allows the vessel to pitch and to roll. Further, this mooring system has, in most of the cases, a ballast tank which is filled with the water so as to provide necessary restoring force to minimize the vessel motion.

This mooring system ensures the product transfer through jumping hoses thereby eliminating the requirement of a submarine hose. Further, maintenance is comparatively easy here as almost all the mechanical equipments are located above the sea level.

Tower mooring system is the cost effective and reliable mooring option for permanently mooring the ships/FPSOs/FSOs in shallow as well as medium water depths (say around 300 ft) where the depth as well as the high currents make external turret mooring option not an easy option to go for.

2.2 MOORING ACCESSORIES

In order to understand the moorings, its schemes and its patterns, the understanding of various mooring accessories are required. The list of such mooring accessories is given below:

2.2.1 Mooring Rope

2.2.2 Mooring Chain

2.2.3 Shackles

2.2.4 Pendants

2.2.5 Swivels

2.2.6 Pyramid Anchor

2.2.7 Mushroom Anchors

2.2.8 Marker and Regulatory Buoy

2.2.1 Mooring Rope

Mooring ropes are used for heavy weather mooring. These ropes are made of nylon, polyester, polyester-nylon mix, polypropylene etc, are multi-stranded and are specified as per the given diameter and strength. Premium Nylon Rope is single strand, double strand or 3 strand and is strong, durable, flexible and shock absorbent. It is a good line for moorings, anchor line, dock lines and many other uses.

2.2.2 Mooring Chain

The use of a heavy bottom chain adds strength to mooring and reduces the angle of pull on the anchor itself. The weight of this mooring chain helps to absorb the shock by lifting slightly when the boat pulls back. A swivel between the top and bottom mooring chains let the boat swing without kinking the chain. A smaller top chain goes from the bottom chain to the mooring ball and creates an arc when the boat pulls back. These mooring chains are hot galvanized grade 43 long chains.

2.2.3 Shackles

All shackles are hot galvanized, forged and heat treated with size and working load marked on each. Average breaking strength of shackles is normally 6 times working load limit. All shackles are safety wired (stainless

steel or nylon Tie Wraps) before installing mooring to prevent them from working loose. It is good to use the same size shackle as chain size for all connections.

2.2.4 Pendants

Pendants are used to connect the boat to the mooring buoy. All Pendants are made with a 3 strand premium nylon marine rope with chafe gear. They are made with a heavy duty galvanized thimble on one end and an eye splice on the other with chafe gear for added protection.

2.2.5 Swivels

Swivels are used to prevent mooring chain from snarling. It is normally placed between top and bottom chains. These swivels are made of hot galvanized and forged steel and tempered. It is of two types: Eye-to-Eye Swivel and

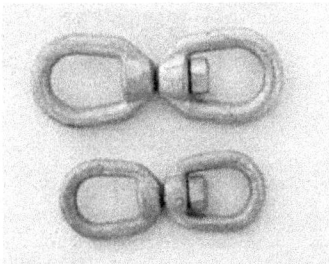

One-Piece Swivel called Shackle-to-Shackle swivel. This shackle to shackle swivel is preferred now a day because a chain assembly is always rated to the weakest link/component and in salt water; the components tend to need replacing before the chain. This One piece swivel is hot dip galvanized and drop forged.

2.2.6 Pyramid Anchor

Pyramid anchor is used in shallow water and it gives more protection in shallow water with less chance of chain wrap-up. This works well in both fresh and saltwater. These pyramids are designed for quick setting. These kinds of anchors are used for boats, docks or buoys.

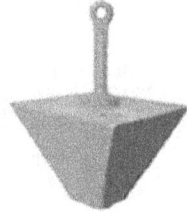

2.2.7 Mushroom Anchors

Mushroom anchors are used for permanent moorings since a long time back. Mushroom anchor name comes from the shape of anchor that resembles that of a mushroom. These types of anchor, normally made of steel, have the advantage of more counter weighting, a more aggressive biting edge, and a larger bell diameter that greatly improves holding power.

2.2.8 Marker and Regulatory Buoy

Marker buoys are used to mark swim areas, No Wake zones, Hazards, and other dangers. Marker Buoy offers high performance design, stability and high visibility. Regulatory Buoy is made out of white polyethylene with closed cell foam inside. Both these buoy has got labels. Labels are fade resistant and highly visible and are sold in pairs. Normally two labels are required per buoy.

We also have solar powered lantern kits that fit both the marker and regulatory buoy. These kits are made with a high intensity LED light. They

provide one nautical mile of visibility. These light turns on automatically in darkness and recharges in daylight. Battery normally last four to five years. Light unit is completely sealed to be moisture proof and shock resistant. The unit should be completely waterproof, vibration-proof and vandal-resistant. It should be designed for autonomous operation without any maintenance (i.e. bulb or battery replacement) for substantial good period of time say 4 to 5 years.

Floating Production System: Process Facilities and Systems

Journey of well fluid begins at well bore. The fluid flow from well bore at sea bed reaches the processing facilities at the floating production system via sub-sea templates, sub-sea pipelines and through rigid or semi-rigid or flexible risers. As the well fluid rises in well bore and travels toward surface, the temperature and pressure falls. Due to these pressure and temperature variations, the fluid separates into three phases. On the Floating Production System, these well fluid are separated into three phases using the on-board processing facilities with the aim to maximize oil production and meet desired product specifications.

The processed oil is stored in the hull of the FPS and gets regularly offloaded to the shuttle tankers via offloading lines and by using buoy mooring. The associated gas so separated is compressed further and then dehydrated to meet the dew-point specifications before its being used either as fuel gas or as re-injection gas for gas lift purposes. Till date, no FPS has the gas dispatch facilities, so gas dispatch or transportation by FPS is normally not there. Produced water gets conditioned on board and then it gets injected back into the reservoir for pressure maintenance purposes or it gets dumped overboard after meeting the requisite effluent disposal specifications. A schematic process diagram below gives a flow diagram for the processing facilities on-board a floating production system.

The process systems on Floating Production Systems mainly comprises of the following, in modularized form:

3.1 Risers and Flow-lines module
3.2 Processing module: Separation systems (Separators and Well Fluid Heat Exchangers)
    3.2.1 Oil module
    3.2.2 Gas module
    3.2.3 Produced Water module
    3.2.4 Crude Transfer Pump modules
    3.2.5 Process Gas Compressors modules

3.3 Sour Gas Processing System (if FPS deployed in sour gas fields)

3.4 Venting and Gas Flaring systems/modules

Plus associated Valves, instrumentations, accessories and controls

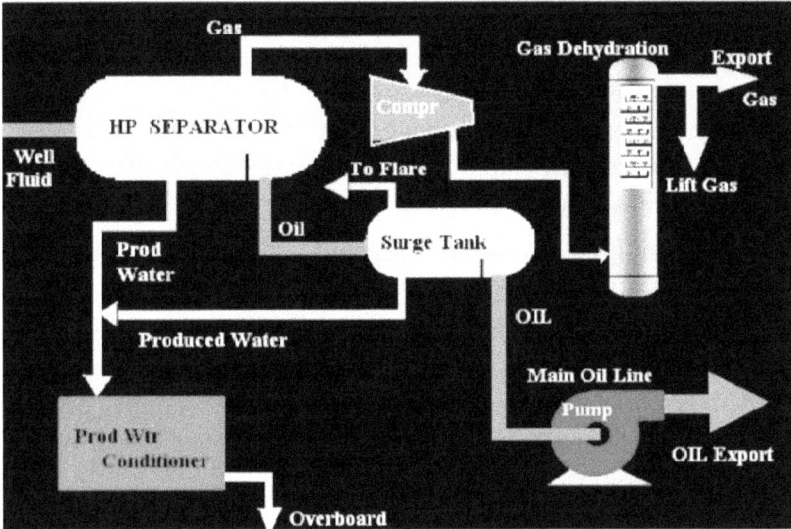

Process Schematic Diagram

3.1 RISERS AND FLOW LINES MODULES

Oil processing facilities at the floating production system are connected with the sub-sea production system through rigid or semi-rigid or flexible risers. See "the sub-sea flow lines and riser" paragraph at chapter 7, sub-sea production system, to understand this riser and flow lines as comprehensive whole. Sometime, we have multiple flexible risers that are used to connect (through multiple connectors) the multiple sub-sea well head located on the sea bed to the processing facilities on the deck of the floating production system. And while doing so, the help of the buoy is taken to ascertain its position and configuration. The riser design concept is quite important here owing to the technology schedule and cost factors. Hence, a number of studies, diagnosis, and fatigue-analysis are required to be carried out beforehand taking into consideration all the sea-data, environmental data, reservoir, field and production data as well. Riser configuration is equally important. So it is important to undertake a static analysis of various riser configurations using the appropriate software.

Risers and Flow Lines can be either flexible pipe or steel pipe. Normally, Flexible pipes are selected over steel pipe for the flow line material, as flexible pipes has:

- Lower installed cost compared with double insulated rigid steel pipe
- Ease of installation and recovery, and can be reused
- Good insulation property, as the crude transported is of a high pour point nature.

The inlet riser at the FPS has a non-return valve and shutdown valve. It is also equipped with pressure switch high low for safety. The pressure and temperature of the line is continuously monitored remotely through automated flow recorders. This helps in identifying if any upset condition has taken place in the wells sub-sea. The shutdown safety systems of the FPS ensure tight shut of the shutdown valves in case of abnormal process conditions.

The flow-lines (oil, gas or water) are interconnected by means of pigging module, which is placed there on the outboard templates allowing the round trip pigging for commissioning, cleaning and inspection purposes, thereby maintaining their the health and integrity. Pig is launched from inlet and received at the outlet thereby cleaning the pipeline by scrapping action. This pigging module is normally remotely operated pigging valve, a valve that also isolates the two headers.

3.2 PROCESSING MODULE

Processing module at any floating production system comprises of the following individual modules:

- Oil Module
- Gas Module
- Produced Water Module

Before we discuss each of these modules in some details, let's understand the very basics of oil and gas processing system, i.e., Separation system comprising of Separators (High Pressure and Low Pressures), Surge Tanks and Well Fluid Heat Exchangers.

We can refer to either of Tables 2, 3, or 4 to understand the various equipment used aboard any floating production system.

Separation System (Separators and Well Fluid Heat Exchangers)

Separators are field vessels used to separate the crude oil coming from oil wells into three components viz. oil, gas and water. This separator is the heart of any oil processing system in oil and gas industry. Hence, It is very important to understand the basic principles of separator functioning and its components.

The components of a separator are as follows:

Primary Separation Section

This section collects and removes the bulk of the liquid in the inlet stream. This is done by changing the momentum of the inlet stream either by creating a centrifugal force or abrupt change in the direction.

Secondary Section

The gas velocity coming to this section is such that the entrained liquid droplets can settle by gravity. Internal baffling is used to dissipate foams, reduce turbulence and accelerate liquid drop removal.

Mist Extraction Section

This consists of a series of vanes/woven wire mash pad and this section removes small droplets from the gas stream.

Liquid Sump

These section collets liquid separated from gas and also provide sufficient capacity to handle surges. Adequate retention time is required for better separation.

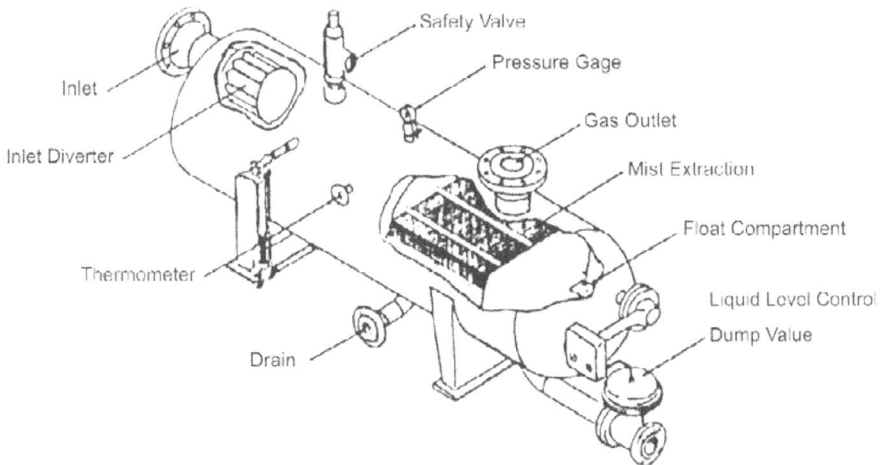

Schematic diagram of a Horizontal Separator

The kinetic energy of the inlet stream is dissipated by the deflection baffle. Predominantly the gas phase flows around the deflection baffle enters the secondary setting stage, while the separated liquid and any entrained gas falls downward. In the secondary setting stage, remaining liquid particles are removed by lowering the gas velocity and using mist extractors. Baffles and other internals are used to reduce turbulence in the liquid collection section thus facilitating the rise and escape of the entrained gas bubbles. Flow distributors such as weirs, plates or vortex baffles are often used to minimize gas re-entrainment in the liquids, which are ultimately withdrawn from the vessel. Whenever crude properties so necessitates, additional treating equipment like crude oil heating, desalting, sand and sulpher removal, heavy oil dilutions and so on are also used besides doing with increasing the number of stages of separation.

Normally, the processing modules have two separator trains for processing: high pressure separators and low pressure separators. The processing module may also have a surge tank that normally operates at atmospheric pressure or at very low pressure say of 1–2 kg/cm^2 pressure and which acts as a storage or as a collector for oil transfer by crude oil pumps. The separators are controlled through individual local and/or control room panel. Normally, the separators on FPS are horizontal three-phase separator. The fluid enters the separator through an inlet shutdown valve, Oil, gas and water are separated in the separator and three phases follow three different paths till their designated final destinations.

Separators are of two types: Vertical and Horizontal. However, Horizontal Separators are preferred because of the following advantages:

- Requires smaller diameter for a given gas capacity
- Can be skid mounted and shipped easily
- Counter flow of gas does not oppose drainage of mist extractor
- Large liquid surface area for foam dispersion
- Can reduce turbulence.

Also, please refer to Table 1 to understand how a given separator is specified for a given process requirement.

3.2.1 Oil Path or Oil Module

Separated Oil from High Pressure (HP) separator goes to Low Pressure (LP) separators via heat exchangers for next phase of processing (LP separator may or may not be there) and thereafter oil is fed to the surge tanks via heaters for further lowering of water content. Surge tank is again a separator with similar construction. Surge tanks can operate in two modes: in "pressurized mode" surge tanks are operated at 4–5 kg/cm^2. It's used when oil is to be sent via pipelines. Operating at this pressure helps in reducing loss of lighter fractions into flare. When the oil is to be sent to tankers it has to be stabilized, in other words all the gases are to be knocked off. In this case the surge tanks are operated at atmospheric pressure. It's evident that there are bound to be stabilization losses in this process. Stabilized crude gets offloaded from the floating production systems to shuttle barges or tankers. The rate of crude transfer is dependent on the size of the shuttle tanker and the loading time. Custody transfer of crude can be done in two ways: tank gauging and positive displacement meters. Metering has an accuracy of $\pm 0.025\%$ of total capacity, while automatic tank gauging, with a servo-operated system, has a reported accuracy of ± 1 mm of the actual level. Centralized Control Panel (CCP) has a flow rate integrator to know quantity of oil dispatched.

Normally most of the produced water gets removed from the crude by the three-phase separators. However, it is anticipated that the residual water which is not separated from the oil, if any, will settle down in the storage tanks in view of the long retention time (4–5 days) available before offloading, and in case the temperature is relatively high, the temperature will enhance further separation. The oil separated here is routed back to LP separator or surge tank. If an emulsion problem occurs, consideration should be given to installing an electrostatic treater or chemical injection facility. This possibility, however, is to be considered in the engineering phase. An exact treatment procedure has to be tried out with crude samples, under laboratory conditions.

3.2.2 Gas Path or Gas Module

After initial separation, the high pressure gas is fed to the compression system by a common manifold called compressor inlet manifold from where gas either goes for delivery (normally with FPS, this gas delivery option for use by external customers are not available) or for getting used as fuel gas (for running turbine, compressors) or being used as re-injection gas for gas lift purposes (gas is injected back into the reservoir so as to enhance oil production). But prior to the delivery or re-injection, this HP gas passes through dehydration process in a dehydration module wherein dew points of the gas is maintained to a specified level by stripping heavy components from gas. Low-pressure gases LP separators or from surge tanks are flared. Gas is measured at outlet of separator using an orifice meter. Orifice in the gas line gives the gas flow-rate. The CCP also has a gas flow integrator. Further, if we have to handle sour gas (H_2S) or acid gas (CO_2), some additional treating systems or modules are required. This will be discussed in some details ahead in this chapter.

The gas processing system comprises of the following:
- Gas manifold
- KOD and piping

- Compressor
- Dehydration skid with Glycol regeneration
- Metering

The gas processing goes like this: The separated gas from the separators (HP and LP) and surge tanks is routed to the compressor package through the gas manifold. Thereafter the gas is routed through the suction KOD and the liquids are scrubbed out. The gas from the suction KOD enters the first stage of the compressors through a strainer. The compressed gas is cooled by air-cooled coolers. In the cooler the gas passes through parallel finned tubes. Air is forced through the tubes with fixed blade fans run by electric motor. The temperature is controlled by changing the angle of guide vanes installed above the finned tube bundles. The cooled gas passes through the first stage KOD. Therefore it enters the second stage of compressor and cooler. Again it passes through the second stage KOD, third stage compressor and cooler. Thereafter, the compressed gas which is saturated with water vapor is fed in to a gas dehydration system wherein the moisture from the gas and other heavy particles gets removed by the hygroscopic action of the tri-ethylene glycol used in the dehydration system. The dehydration system is designed to give a specified dew-point for the gas before the gas being put in use. Further, as an added preventive measure, gas corrosion inhibitor is also added before the compressed gas gets ready for dispatch or gets ready for further use. The gas is also metered continuously. For the purpose of metering, the gas passes through an orifice plate and the differential pressure, static pressure and temperature are measured. These parameters are processed into electronic signals and transmitted to the flow integrators and recorders in the control room.

Gas Dehydration Module

To prevent corrosion, hydrate formation and condensation of water, the gas needs to be dehydrated before its transportation for gas lift purposes or for its use elsewhere. There are many chemicals having good hygroscopic properties that are used to strip heavy components from gas, thereby maintaining the specified dew-point for gases. Normally TEG (tri-ethylene glycol) is used for its better absorption capacity.

This compressed gas, which is saturated with water vapor, is fed in the glycol contactor of the gas dehydration module. The glycol contactor has two sections. The gas enters from the bottom and passes through the scrubbing section wherein mist extractor removes the liquid droplets. Thereafter it enters the other section wherein the lean glycol (TEG, tri-ethylene glycol) falls from the top gradually and the gas either passes through the valve tray or bubble cap tray. Normally there are six to eight trays. The gas and glycol comes in intimate contact over these trays wherein the lean glycol absorbs water. The dry gas devoid of moistures comes out of the contactor. Under normal operating conditions, the concentration of TEG 99% by weight at contractor inlet top tray and 95% by weight at contractor outlet after dehydration has taken place. Thereafter it passes through a gas/glycol exchanger where the high temperature of glycol aids in increasing gas temperature. Thereafter the gas is sent to the reflux condenser which is also an exchanger and the water vapor from the glycol still again increases the temperature of the gas and the water thus condensed is sent to the reflux drum and to the still as a reflux. The gas thereafter passes through an overhead scrubber where the liquid droplets, if any, are scrubbed and the dry gas is dispatched for its designated uses. The pressure in the contractor is controlled through a PCV located downstream of the overhead scrubber.

The rich glycol comes out of the contactor and is sent to the glycol flash drum to remove dissolved hydrocarbon. The gas thus released is sent to the LP flare KOD. The glycol is skimmed off manually. To remove any solid particles and liquid hydrocarbon, the rich glycol is thereafter passed through the cartridge and active carbon filters. Thereafter the temperature of the rich glycol is increased by passing it through a glycol/glycol exchanger, wherein the hot lean glycol helps in increasing the temperatures of the rich glycol.

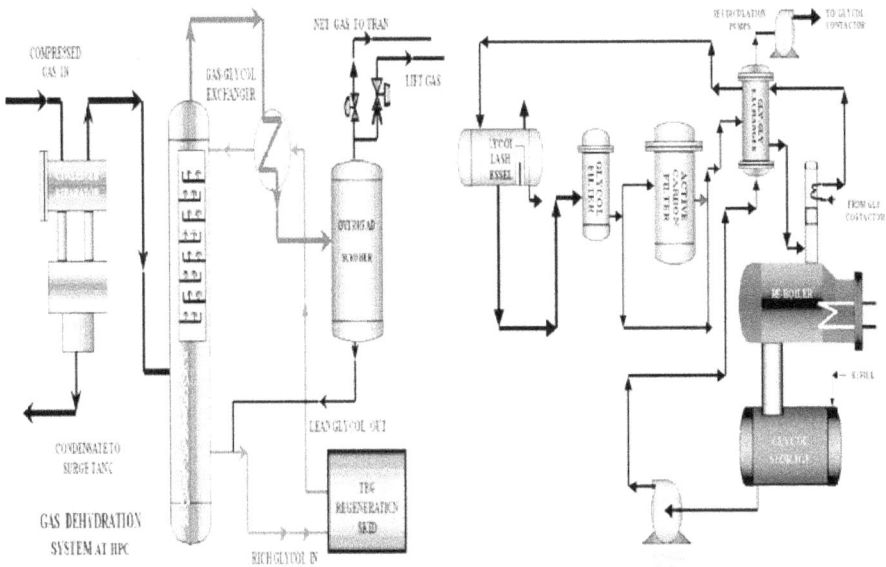

Gas Dehydration and Glycol Regeneration Schematic

Thereafter the rich glycol enters the still of the re-boiler. The re-boiler is an atmospheric vessel and operates at about 200°C. The glycol is heated with the help of hot oil. The hot water vapor leaves the top of the still and is sent to the reflux condenser. A part of the condensed water is sent to the still as a reflux from the reflux drum. The lean glycol overflows a weir and drops to the storage tank through a packed column. In order to achieve higher glycol concentration (99%) stripping gas is used in the storage tank. This gas comes out along with the water vapor and gets vented. The stripping gas reduces the partial pressure of water vapor and improves the glycol concentration.

The regenerated glycol is stored in the storage tank and is pumped to the glycol/glycol exchanger to reduce temperature through a booster pump. Thereafter the re-circulation pump increases the pressure and the lean glycol is sent to the glycol contactor tower. Before entering the tower it again passes through a gas/glycol exchanger wherein the temperature of the lean glycol further drops and the temperature of the gas for use increase.

To see whether gas has got its specified dew point or not, the dryness of the gas is measured by passing it through a moisture sensor. The capacitance type probe has got active alumina as the dielectric medium. The dielectric constant changes with the presence of moisture in the gas and gives a value of the dew-point of the gas.

3.2.3 Produced Water Path or Produced Water Conditioning Module

The water separated in the separators and surge tanks are sent to the produced water conditioner wherein the oil is skimmed off and the treated water is either dumped in the sea or in the sump caisson. Sometimes, this produced water is also used for water injection. And accordingly, this produced water passes through a series of treating, de-oxygenation and conditioning process prior to injection.

Environmental and state/national regulations requires that produced water from the separators be treated to reduce the oil content in the produced water to a specified level, say 35 ppm, or less, before it can be safely disposed to the sea. The oil quantity in water to be disposed at sea depends upon the region in which FPS are operating as different regions have their different guidelines w.r.t. the maximum contamination of effluents like North Sea (25 ppm), Canada (25 ppm), China (15 ppm), S.E. Asia (50 ppm). For Indian Offshore water, it has to be less than 50 ppm.

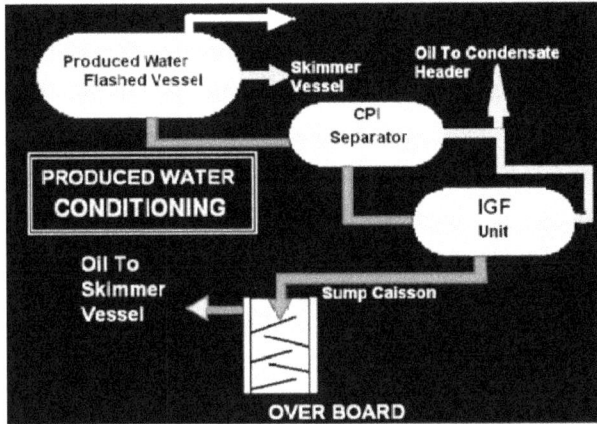

Produced Water Conditioning Process

Different processes are available for produced water conditioning or for effluent oil water separation like Gravity Separation; Floatation; Coalescing; and the recent one being Hydro-cyclone.

Gravity Separation

Gravity Separation is the simple process and the most common one. It creates a quiescent condition, under which the buoyancy force raises the dispersed droplets to the surface for skimming. Gravity separation comprises of Skimmer vessel, CPI separator and Parallel Plate Separator (PPS). The PPS consists of titled parallel plates. The effluent passes through the plates and free oil particles raise till it reaches the underside of the plate, gets collected and coalesced. The coalesced oil particle travels upward till it reaches the water surface wherein it is skimmed off and collected. PPS further are of two types: Cross Flow Interceptor (CFI) and Tilted or Corrugated Plate Interceptor (TPI). The CFI comprises of flat plates tilted at 45° and forms a series of chevrons. The TPI/CPI utilizes a stack of parallel-corrugated plates at an angle of 45°.

Floatation

We have two types of floatation: Dissolved gas floatation and Dispersed gas floatation. In dispersed system, finely dispersed gas bubbles are introduced into the produced water stream entering into the system. Here, the increase in the rising velocity of oil droplets is achieved by increasing the differential density by attaching gas bubbles to oil droplets. The system has provision to skim floated oil particles. In case of dissolved gas floatation, gas is dissolved in produced water under pressure. In the floatation cell the pressure falls and the gas bubbles come out entraining oil droplets. The IGF uses rotors or educators for entertainment of fairly coarse bubbles in the carrier stream.

Coalescer

Coalescer provides sufficient surface area for the oil droplets to coagulate. Normally granular materials like sand, anthracitic, resins or a bundle of porous tubes are used as the coalescing medium. It can be up flow or downfall type. We have two types of Coalescer: Vertical tube coalascer and Packed bed coalascer.

Here, produced water enters the conditioning system or module at the plate coalesce for further removal of suspended oil prior to discharge into the produced water tank for setting and later disposal to sea. Plate coalescers are recommended if one want oil content of about 35 ppm or less. This system offers excellent water qualities with 98% removal of oil particles greater than 50 microns in size, and oil content of less than 35 ppm from discharged water. Further this system provides high reliability in operations as there are no moving parts, no electrical power is required and also it has low operating cost. However, if the requirements call for a maximum oil content of 20 ppm or less, a flotation cell should be considered.

Hydro-Cyclone

It is based on the centrifugal force principle, where the force including the separation is much more than that achieved in the gravity separator. The fluid enters the hydro cyclone, spin is imparted and a very high centrifugal force is generated, which causes practically instantaneous separation. For an effluent with very high oil PPM, the hydro cyclone may be answer. The produced water reaches the conditioner through the produced water header.

The two important equipment in processing Oil and Gas systems are "Main Oil Line Pump" placed in the oil path post separation and "Process Gas

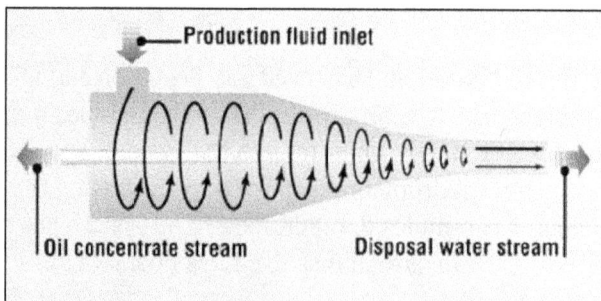

Production fluid inlet

Oil concentrate stream   Disposal water stream

Compressors" placed in the Gas path post separation. It is important to develop an understanding of these two "Critical Equipment"; critical because its non-availability causes a total system shutdown, if standby are not available.

3.2.4 Main Oil Line Pumps (MOL) Module

The main oil line pumps are horizontal, split case, multistage skid mounted unit which is used to transfer the crude oil either to the hull of the tanker or for offloading crude oil to the shuttle tankers. These pumps are equipped with most advanced controls system and operating mechanisms. The pumps are having inlet and outlet SDV, which are motor operated and known as MOV. The electric motor is directly coupled with the pump shaft whereas the gas generator is coupled to the gear – box and thereafter to the pump shaft.

Let's understand the operation of MOL. By pressing the start button from the local panel or control room, the auxiliary lube oil pump and cooler starts up. It develops some pressure and at the predetermined value, the MOL pumps will start. The suction MOV gets opened earlier and the discharge MOV starts opening after sometime. The flow across the pumps is controlled by the LCV, downstream of the MOL pump. A recycle line maintains the flow across the pumps in case the level is low in the surge tank. The main lube oil pump is mounted on the same shaft of the MOL pump. After the lube oil pressure reaches some predetermined value the auxiliary lube oil pump cuts off. When the pump is stopped, then immediately the auxiliary lube oil pump will start and the discharge and suction valves will get closed. The lube oil system will stop after some predetermined time. The high pressure crude passes through the turbine meters, which measures the quantity of crude being pumped to the storage and transportation facilities attached with any given FPS.

Normally, these MOL pumps have a standby provision; a provision by which the standby pump can takeover automatically in case of failure of one pump. Apart from the vibration probes, the pumps are fitted with PSHH (discharge), PSLL (suction) PDSL (differential), PSL (lube oil), LSL (lube oil), and TSH (motor bearing) for operating the pump safely.

3.2.5 Process Gas Compressors (PGC) Module

The process gas compressors are used to compress the separated gas to a specific discharge pressure for either its transmission or for its use somewhere, internal or external to the FPS.

Let's understand how these compressors get operated. The compressor essentially consists of a high pressure casing which houses the diffuser vanes and a shaft with series of impellers fixed on it. The casing is of horizontally split type and housed in a high-pressure barrel. The shaft mounted with impellers is balanced statically and dynamically. The material of construction is chosen depending on the composition of the gas to be handled. Labyrinth type seals are used to provide sealing between the impellers, which stops gas leakage from the high pressure to low-pressure side. The shaft ends are provided with liquid seals, wherein seal oil system provides the required sealing.

The compressors are provided with Gas Generators (GG) as drivers. Due to its lightweight and because of wide range of operation and usage of natural gas as fuel, makes GG a popular choice. GG are placed in a closed enclosure. The air is sucked by the axial compressor through intake air filters. The air intake filtration normally has three stages consisting of vaned separator, coalescer and again vaned separator. Thereafter the air passes through a bank of silencing units housed in the plenum chamber for an even flow of air to the GG intake. The quantity of air required varies with the GG speed and is adjusted by the variable inlet guide vanes. This compressed air is sent to the combustion chamber. The fuel gas is controlled through governor and is sent to the fuel manifold where there are set of nozzles. The gas is ignited by an electric start. Normally there are eight combustion chambers placed equidistantly at the downstream of the compressor. The hot gas leaving combustion chamber enters the turbine section (2 or 3 stage) of the GG, which drives the shaft of gas generator.

The fuel gas after leaving the gas generator turbine enters the single or double stage power turbine generating required speed and power before it is discharged in the atmosphere the starter motor is used to start the gas generator. It is connected to the axial compressor shaft through turbo-flex coupling, ratchet housing and ratchet pawl. Ratchet pawls are spring loaded to engage at low speeds and counter weighted to disengage at high speeds.

The output of the power turbine shaft is connected to a speed increasing gear box, which in turn is connected ton the compressor shaft with solid coupling. The turbine enclosure is ventilated with atmospheric air. Two sets of blower fans are used, one for blowing in and other as exhaust; a negative pressure is maintained to ensure proper ventilation. The control panel accommodates all important operating parameters, machine status and control switches. The panels of both the GG and compressor are located in close proximity. These also have local panels located in the skid. The pressure, temperature, vibration, axial displacement, fire and gas safety status are continuously monitored and displayed.

In process gas compressors, we usually face the problems of "Surging". Surging may be described as an unstable flow due to pulsating action of gas, which occurs when the differential pressure between the suction and discharge reaches a certain level. This results in gas hammering within the compressor and can damage bearing, seals, impeller; shafts etc. of the compressor if it continues for a longer duration and can cause extensive fatigue. Surging can be prevented by a surge control system. A part of the gas is re-circulated back to the suction. The sequence and quantity of re-circulation is computed and executed through a programmable controller, which operates a control valve. The quantity of gas depends on the head/flow status relative to the machine speed.

3.3 SOUR GAS PROCESSING SYSTEM OR MODULE

If the Floating Production System is deployed to exploit and produce a marginal gas field; gas having sour properties, then FPS must have in its fold a separate module comprising of "Sour Gas Processing" facilities.

Sour Gas Processing comprises of following components:

3.3.1 Gas Cooling
3.3.1 Separation
3.3.1 Gas Dehydration and Glycol Regeneration
3.3.1 Condensate Handling
3.3.1 Desulphurization Process

3.3.1 Gas Cooling

The gas coming from the wells is at high temperature and is therefore being cooled by coolant water in shell and tube heat exchanger before sending it to separator. On tube side there is gas and shell side there is water. On outlet of gas, there is temperature switch installed for unit shutdown in case the temperature of gas exceeds.

3.3.2 Separation

During the normal production, gas from the well is sent to separator through gas production header. The pressure in the vessel is maintained at constant value by discharging the gas from the separator by means of pressure control loop installed on the gas outlet line. The separation of condensate, water occurs in separator with formation of two liquid phases, which are discharged separately under level control. The condensate/oil level is maintained by discharging the condensate/oil through level control valve to the surge vessel. The water level is maintained by discharging the water through interface level control valve to produce water system.

3.3.3 Gas Dehydration and Glycol Regeneration

The gas coming out of separator and after getting compressed by compressors must pass through gas dehydration system to have a dry gas having devoid of any moisture and having a specified dew-point. The basic operation of the glycol dehydration remains the same as discussed previously in this chapter.

3.3.4 Condensate Handling

Condensate handling system consists of a surge vessel and a coalescer. In surge vessel, water content in the condensate is reduced to a maximum of 3% and in coalescer; water content is reduced to 20 ppm by volume. During the normal production, the condensate from separators and test separator is sent to surge vessel. Due to pressure decrease between gas production separator, test separator and surge vessel, partially flashing of condensate occurs. Gas is sent under pressure control loop installed on gas outlet line to fuel gas system in a split range. The separation of condensate and water occur in the separator with the formation of two liquid phases, which are discharged separately under level control. The condensate level is maintained by recycle line and water level is maintained by discharging the water through level control. Condensate from vessel is then sent to coalescer where little water in condensate is reduced to less than 20 ppm. Coalsecer is a horizontal vessel and the condensate flows in it and exits without level control to condensate metering. The water level is maintained by discharging the water through level control valve to produced water system.

3.3.5 Desulphurization Process

Desulphurization is a very important process at any Floating Production System for removing H2S from the gas to be utilized at FPS. It forms a very important part if we are using FPS in a sour gas field. Hence, it is imperative to understand in some detail this desulphurization process.

There are three desulphurization processes:

- Amine-absorption Process
- Iron-Sponge Process
- Molecular Sieve Process

These process are explained in detail at Annexure G.

3.4 VENTING AND GAS FLARING SYSTEMS OR MODULE

Normally gas is used for fuel as much as possible and excess gas gets disposed to flare. The HP gas may be used as fuel gas for gas turbines and diesel engines; the LP gas may be used in boilers and other miscellaneous equipment. Excess HP and LP gases are collected or flared. Equipment which uses the available gas is Gas turbine drivers for power generation or pumps; Inert gas generator; Pilot gas for flare; Boilers for steam generation. It is also used in Gas lift and gets re-injection into reservoir, or gets transported to shore through a pipeline.

In any flaring system, considerations must be given to have a safe radiation level and also to avoid the liquid carryover. Special attention shall be given to prevent condensate dropout, liquid carryover, and hydrate formation in the flare system. For personal safety, a maximum radiation level of 250 Btu/hr/ft^2 is normally considered a safe option. Further, disposal of gas with high hydrogen sulfide content can be hazardous. So before disposing high H_2S gas to flare, it has to undergo desulphurization process.

3.4.1 Vent and Gas Flaring Types

Vent and gas flaring are of three types:

- Atmospheric vents
- High pressure gas flaring
- Low pressure gas flaring

The vent header is atmospheric. The gas from the receivers, launchers, separators (depressurizing), surge tank (depressurizing), fuel gas conditioning (depressurizing), process water separator, skimmer vessel, sump caisson etc., can be released through the vent boom. The gas passes through a KOD before it is vented. The vent booms are equipped with CO_2 snuffing unit to extinguish fore. Some of the vessels like chemical storage tanks; diesel storage tanks have independent vents. The process gas compressor packages also have independent vents to depressurize the system.

Apart from the cold vent there are two headers, high pressure and low pressure, through which the gas is sent to flare. The high-pressure vessels like separator, surge tank, compressors, dehydration skid, fuel gas skid etc, are connected to the high-pressure flare header. The liquid carried over by the gas gets knocked off in the HP flare KOD before it is sent to the flare. The liquids thus drained from this KOD goes to the closed drain header. The relief valves, depressurizing of process equipment and occasional manual venting of process equipment can be done through the high-pressure system.

The low-pressure system also has flare KOD. Gas from the surge tanks during stabilized mode and glycol flash drum is sent to the flare through this system. The liquid drained gets collected in the skimmer vessel. Since the flow through this header is normally very less, therefore this header is purged continuously with fuel gas.

3.4.2 Flaring Types

The objective of the flare is to burn the combustible gases. Normally there are three types of flares:

- Pipe flaring
- Tulip flare
- Fin flare

The pipe flares are simple in construction and is just an open-ended pipe.

Normally these are equipped with windshield, pilot burners and thermocouples (to measure temperature). The high-pressure gas comes out from the center pipe and low-pressure gas through the annulus.

The tulip flare utilizes a skin adhesion affect known as 'coanda effect', the principle of which is simple: when high-pressure gas is ejected from the annular slot, it changes direction and follows the profile of the flare tulip entraining air. The gas entrains sufficient air approximately at the point of maximum diameter of the tulip for combustion to begin at the outside of the gas film. Combustion takes place from outside to inwards, thus there is always a protective layer of unburned gas between the flame and the tulip preventing flame impingement on the tulip metal. At lower flow rate due to subsonic flow, the combustion characteristics approach those of a conventional pipe flare wherein the desired cooling effect reduces on the flare metal surface.

The fin type is like pipe flare, but instead of the gas coming out of the pipe; it is routed through holes in the fins welded around the circumstances of the tip. This is meant for better combustion. But in case, if liquid hydrocarbon is carried over then it will cause high flare temperature and can damage the fins.

To avoid liquid going to the flare, a high liquid level switch is provided in the KOD, which when actuates causes shutdown of the processing system. The high pressure and low-pressure flare lines are interconnected with valves and bursting disc. In case the pressure becomes excessive then the disc ruptures and the pressure gets released through the other header. The flares are also equipped with pilot burners. A flame front generator or Pallet Launching Unit (PLU) is provided on the FPS to ignite the flare if it gets extinguished.

3.4.3 Flaring System Alternatives at Floating Production System

On a given floating production system, there exist various alternatives for gas disposal like:

- Remote flaring
- Onboard incinerator
- Onboard conventional flare and boom
- Ground flare

Remote Flaring

This flaring is attractive in some cases. Because of its remoteness from the storage and production facilities, the potential fire hazards and health hazards are reduced. It is desirable to have the wind blow the hazardous gases away from the production facilities and personnel. At a minimum distance of 500 meters, the gas will be sufficiently diluted to be non-hazardous even if the wind direction is reversed. In shallow waters, the remote flare may be mounted on a fixed structure. As the water depth increases, the cost of a fixed structure will be very high. As such, a buoy might be a better solution. However, a buoy will be suitable only for mild environmental conditions. In deep waters, the erection of a fixed structure is not economical, whereas a flare buoy is not adequate to survive the harsh environment. Hence, remote flaring is not suitable at deep waters. For the barge FPS, the water depth is relatively shallow and a remote flare mounted on a tripod is the solution.

The Onboard Incinerator

The onboard incinerator is firebrick lined. It uses about 25% excess air to burn the gas and additional air to quench the exhaust gas. However, this is not preferred because of the dimension it takes and process it utilizes. For example, a unit 24 feet (7.3 meters) in diameter by 50 feet (15 meters) tall, weighing 70 tones, will be required to dispose of 12 MMscf/d. Further, hot firebrick is considered a fire hazard onboard the vessel. Also, incinerator requires for special corrosion-resistant materials, which are costly, and the large volume of cooling water make such an installation impractical for burning gas in large volumes.

Conventional Flare Booms

These are normally installed on fixed offshore platforms and on semi-submersibles. There are, however, major objections to using such flares on tanker-and barge-based FPSO's. The boom length would have to be about 230 feet (70 meters) to ensure an acceptable level of heat radiation on deck. This length may create structural problems during roll conditions. Secondly, an open flame during offloading operations can pose a safety hazard. In the barge-based FPSO case, the barge will be fix-moored. Two cantilever flare booms will be required to account for the changing direction of the wine. A shuttle barge will be moored alongside to offload the crude. The cantilever flare booms will interface with the offloading operation. Hence, it is not preferred.

Ground Flare

A ground flare has found wide acceptance on weathervening FPSO tankers. It is normally mounted vertically on the deck, in a location where it is downwind of the living quarters. It is normally supplied with forced draft air at the base, to provide a short, non-luminous flame. The burner and the flame are completely shrouded by a refractory-lined vertical stack to reduce radiation levels on the deck. The stack has openings around the base to provide natural cooling of the refractory lining.

3.4.4 Preferred Flare System of Different FPS

Tanker

For Tanker type FPS, a ground flare is preferred and it is arranged in such a way that the living quarters are always upwind of it. Since the tanker is to be stern-moored to an SPM, the living quarters are located in the deck house at aft; the ground flare is therefore to be located near the bow.

Barge

For Barge type FPS, in very shallow water, a remote flare is preferred because of safety reasons and low cost. Two sub-sea lines are required, one for the HP gas and one for the LP gas. The flare is to be located approximately 0.5 kilometers from the FPSO barge, downwind of the prevailing wind direction.

Semi-Submersible FPS

For Semi-submersible FPS, It is preferred that the two flare booms installed on the selected floating semi-submersible platform, one opposite the other, be lengthened for the disposal of gas. The booms are to be cantilevered from the side of the platform. It is proposed to offload crude from the platform via a pipeline and buoy located away from the floating platform; the restriction of having open flame near offloading operations does not apply.

Floating Production System: Utilities and Process Support Facilities and Systems Modules

Following utility and process support system and facilities are generally available on any floating production system either as individual modules or few combined together as a single module:

4.1 Power Generation System Module

4.2 Fuel Gas System Module

4.3 Instrument and Utility Air System Module

4.4 Control Panel and Instrumentation System

4.5 Communication System

4.6 Desalination and Potable Water System

4.7 Utility Water System

4.8 Cooling Water System

4.9 Seawater System

4.10 Chemical Injection System

4.11 Crude Oil Washing System

4.12 Purge Gas (Inert Gas) System

4.13 Vapor Return System

4.14 Diesel Fuel System

4.15 Ballast Water Treatment System

4.16 Hot Oil System

4.17 Sewage System

4.18 Kill Facilities

4.19 Helideck

4.20 ATF Refueling System

4.21 Material Handling System

4.22 Heating, Ventilation and Air Conditioning Equipment

4.23 Drain System

4.1 POWER GENERATION SYSTEM MODULE

The FPS is always equipped with sufficient power generations, which supply power to the living quarters, process pumps, air compressors, lighting, battery chargers, etc. Power generation system normally comprises of Gas Turbine Generators, utility and emergency Diesel Generators, Switchgear rooms, UPS (un-interrupted power supply) and is ably supported by a fuel gas system. Normally the power generation is centralized i.e. leaving the process gas compressor; all the units are fed power from the turbine generator. Further, the size of any power generation system depends upon the overall capacity and requirement of process, utility and process support system along with the driver selection of major prime movers. Accordingly, we design the number of turbine generators, its capacity and number and capacities of other components too.

4.1.1 Gas Turbine Generators

This power generation system consists of gas generator, power turbine alternator, TG control panel, switchgear and transformer. Gas Turbine Generators are used either for power generation or as mechanical drives for gas compressors. The generators are of dual fuel type but normally run on gas, which is easily available. These machines are one of the most critical equipment installed on any FPS for operations. A gas turbine consists of two parts namely gas generator and power turbine along with a host of accessories like lube oil system, seal oil system, intake air system, gas conditioning skid and exhaust cooling systems. Gas generator is a consolidated unit of a multistage inlet air compressor and a two-stage turbine. This critical equipment warrants best of maintenance and repair practices using besides the other things, the state of art DGS-digital control systems and vibration-monitoring systems. Voltage and frequency maintenance suiting standard equipment requirements are done.

4.1.2 Utility/Emergency Diesel Generators

In case of shutdown (ESD), the auxiliary or emergency generator takes over. These emergency generators are diesel engine driven and the capacity is normally between 1.0 to 1.2 MW. These diesel generators are installed for black start up of the production systems/equipment in case of power or emergency shutdown of the floating production system due to various reasons. These generators supply power to gas turbine auxiliary motors, instrument air compressors and lighting panels of the FPS.

4.1.3 Switch Gear Room and UPS

Normally, there is two-switch gear room; one is for high voltage (6.6 KV) and the other 415 Volts. The UPS (uninterrupted power supply; a battery bank), during generator shutdown and total power failure, supplies power to the instrument/controls and some emergency lighting. Each UPS has certain defined "supply hours" say, for example, eight hours power supply back-up.

On tanker based FPS, we normally prefer generators with steam turbine drivers. This is preferred because this indicates that the boilers will be operating continuously and exhaust gas is available all the time, so that no separate inert gas generator is required. Dual-fuelled gas turbine driven generators are preferred for the FPSO barge, using either natural gas or diesel fuel, or both. If gas turbine drivers are selected to drive the generators, it is desired that the waste heat be recovered to generate steam in waste heat boilers for heating purposes. The semi-submersible is equipped with sufficient power generation for its application as an FPS. Power requirement for the FPSO barge is about 800 kW, while that for the FPSO tanker and semi-submersible FPS is about 1000 kW. Refer to Tables 2, 3, 4 and Tables 8, 9 at the back of the book to have an understanding through the equipment list for its power requirement.

4.2 FUEL GAS SYSTEM MODULE

The fuel gas system comprises of suction KOD, compressor, heater, scrubber, filters and super heater. The power generation system, process gas

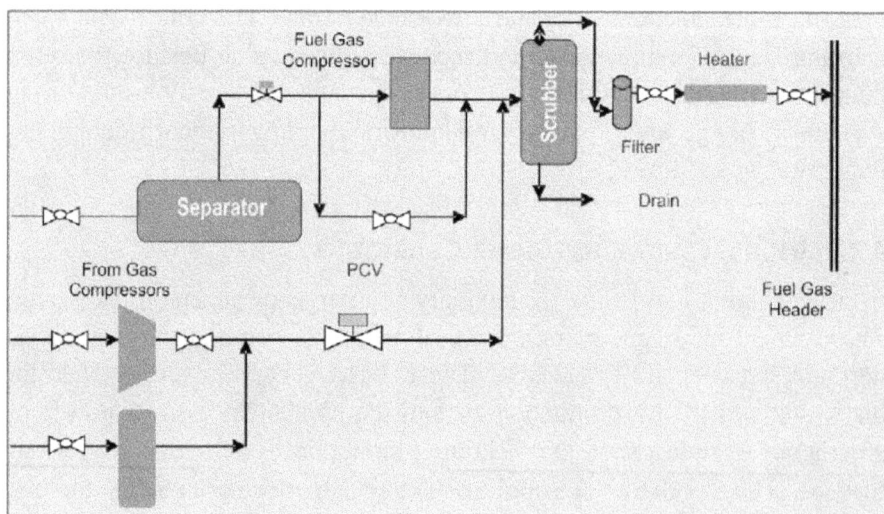

compressor and in some cases main oil line pumps have gas generator as driver. Supply of fuel gas to these generators at a particular pressure and temperature is very vital and is provided by this fuel gas module. Moreover excess liquid should not also be carried over; otherwise this can give hot spit, can choke the injectors and can increase the combustion temperature. Every unit has its own fuel gas valve, which requires supply pressure at some predetermined range, otherwise it can stop supply and the unit will trip. Hence it is essential that the fuel gas be conditioned before it is sent to these gas generators.

Fuel gas can be taped from the separators or from the outlet of the glycol towers. In case, it is from the separators then it needs to be compressed. This is done by the dedicated fuel gas compressor which is a gas engine driven reciprocating type compressor. The gas from the separators is scrubbed by the suction KOD prior to compression. Depending on the requirement, the compressor can be bypassed and by increasing the separator pressure gas can be supplied to the conditioning skid directly. The line from the tower outlet has a PCV, in addition a SDV, being at a much higher pressure. Thereafter the gas is heated up to 55–60°C in the heater. This electric heater is provided with TSH, which will automatically cut off the heater in case of its actuation. Gas from the heater enters the fuel gas scrubber, where the liquid droplets are knocked off. A PCV downstream the heater maintains the pressure in the fuel gas manifold. As a second line of protection, the conditioning skid is equipped with filters downstream of this PCV to take that the gas is absolutely free of liquid. The gas from the filter enters the super-heater wherein the temperature is raised to about 70–75°C.

The fuel gas heater is quite long, even though it is jacketed but the temperature drop can cause condensation. In order to avoid this, super heater is used to increase the temperature. The super-heater also has temperature switch which on actuation will cut off the electrical supply. The individual units have their own toppings from the header and are equipped with SDV and BDV. In case the units shut down, the SDV closes and the BDV gets opened such that the downstream of the SDV gets depressurized. A blow down valve is also provided in the fuel gas header. This actuates in case of ESD or FSD and depressurizes the whole system.

4.3 INSTRUMENT AND UTILITY AIR SYSTEM MODULE

The instrument and utility air system comprises of air compressor, instrument air receiver, after coolers, utility air receiver, pre-filters, and air driers and after filters. All these equipment are placed on an instrument skid. The instrument air skid may typically contain 2 × 100% or 3 × 50% compressors, after coolers, utility air receiver, 2 × 100% heatless air driers, an instrument air receiver and an instrument air controller.

4.3.1 Air Compressors

These compressors are multi-stage reciprocating type and supply compressed air to various pneumatic instruments/panels to control the processing of crude oil/natural gas. The compressed air is passed through air driers to absorb any moisture to ensure proper functioning of pneumatic instruments/panels. Pressure drop in the instrument air header below the prescribed limit will cause shut down of the process system/equipment. Utility air is used to run the pneumatic winches and other pneumatic tools and in surge tanks of service water system. Air compressors are specified by its discharge pressure and discharge flow range. Just to understand, say, discharge pressure range: 7–10 kg/cm^2, Discharge flow range: 500–750 m^3/hr. A dedicated air compressor is also provided for this purpose. These are 'V' type reciprocating compressor, one running and one standby. In case of low pressure, the standby compressor automatically comes into operation. To reduce maintenance expense, capital cost, space, and weight, synthetic oil-lubricated screw air compressors are preferred.

4.3.2 Instrument Air and Utility Air

Instrument air required for operating the process instrument and control valves. The instrument air is used for the instruments (PCV, LCV etc) fire loops, ESD loops etc, and is the most important system of a FPS. Starting air may be required for combustion turbines, for diesel engines, and for alternate liquid fuel atomizing air in the boilers. The utility air is used for the potable water system, utility generator, pedestal crane, operating winches, hose reels at different location on the decks etc. The startup air for the firewater pumps is also supplied by the utility air header. Starting air may be required for combustion turbines, for diesel engines, and for alternate liquid fuel atomizing air in the boilers.

The air from atmosphere is filtered before and after it is compressed. The compressed air is stored in the utility air receiver. From there a part of it goes to the dryer, where solid desiccants like silica gel is used for drying. Normally there are two dryers with one in operation always and it automatically changes after certain number of hours. The unit, which is not in operation, is dried by using electric heaters. Thereafter the instrument air is stored in a vessel called instrument Air Receiver. The flow and pressure in the header is maintained through a PCV. All the pneumatically operated instruments, loops and panels are supplied through this header. In case of loss of pressure, a pressure switch actuates and the standby compressor is automatically loaded. As the Firewater pumps require air for starting, the utility header and the header meant for the start up air are connected. i.e. the utility air can be used for starting up the fire water

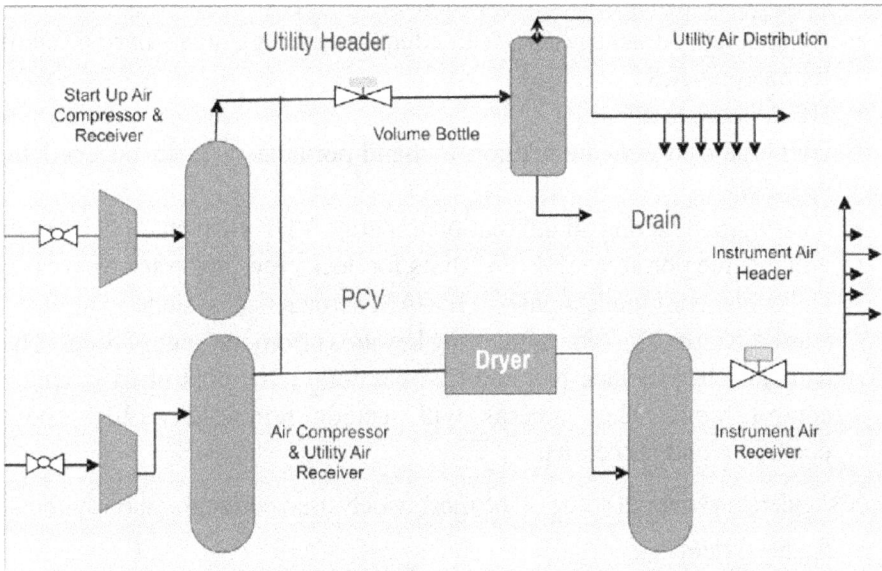

Instrument/utility air system

4.4 CONTROL PANEL AND INSTRUMENTATION SYSTEM

Control room at any FPS does have adequate space for a control panel. The control room will house the process control panel, ESD logic controller, and fire protection annunciation panel. The instruments are normally pneumatic control valves with local controllers. Transmission of alarms, shutdowns, and other out-of-range status signals takes place through 24 volts DC to the control room. All ESD shutdowns get transmitted by 24 volts DC fail-safe signals.

Control room is having Remote Telemetry Unit (RTU) and Remote Platform Monitoring Console (RPMC), which are automatically operated under a pre-designed logic, sequence and software. This remote telemetry unit in combination with remote platform monitoring console assists in operation of well platform from Central Control Room (CCR). It helps in acquisition of data about equipment and wells at CCR.

4.5 COMMUNICATION SYSTEM

The telecommunication system is the direct link to the outside world. It is imperative that reliable equipment is available aboard any FPS, with adequate redundancies to back up failures. The system typically comprises of the following:

- Telephone system approved to tie-in to the local telephone system, and in compliance with the local regulations.
- PA (public address) system with adequate zoning and an alarm system for emergencies
- HF radio system
- UHF radio system in relation to hand-portable units to be used by operators
- VHF system having all international marine VHF channels. The system will include portable VHF-FM units for deck crew and crane drivers
- Helicopter communications—VHF-AM aeronautical system
- Closed-circuit TV—for monitoring key areas from the control room. The cameras shall include pan and tilt operators, to be controlled from the control room. The cameras will include remote-controlled zoom operations and processing.

Let's understand in brief some of the most widely used communication systems:

- Radio system
- Past party communication system
- Closed circuit television system
- Remote Telemetry Unit
- Private Automatic Branch Exchange and Telephone System
- Satellite Phone and P&T lines

4.5.1 Radio System

The radio system comprises of the following:

- Lifeboat radio equipment is to provide for emergency distress communication form lifeboat to a shore radio station or any other life saving agency.
- VHF—FM marine transceiver is to provide FPS to platform and to ocean vessel, communication within the production field.
- VHF—F walkie-talkie sets is to provide operational communication with nearby stations.
- VHF—AM aero transceiver is to provide communication between the FPS and helicopters working in the area.

- Non Directional Beacon (NDB) is to provide a continuous code radio signal to assist helicopters working in the area in locating the offshore installations. HF-SSB transceiver and radio teletype are to provide both voice and teletype communication between platform and shore.

4.5.2 Field or Past Party Communication System

The system is a complete intra FPS as well as an inter FPS page party communication system with an emergency tone generator. The system is common talking, so that any handset user may take part in any conversation. Speakers are used only in the page mode and emergency alarm mode.

4.5.3 Closed Circuit Television System

The Closed Circuit Television System (CCTV) system provides a high-resolution camera system with remote controlled pan, filt and zoom capability and high-resolution monochrome monitors. It is suitable for low light operation. It is useful in monitoring helicopter landing boat operation at any external unknown approach and general surveillance of plant and equipment. CCTV can be linked to any onshore based offshore control room or system for better emergency monitoring by the management.

4.5.4 Private Automatic Branch Exchange and Telephone System

The Private Automatic Branch Exchange (PABX) and telephone system provides an internal switching operating and interfacing system for the telephone network. The system is totally automatic for unattended operation.

4.5.5 Satellite Phone and P&T Phone Line

The satellite phone and P&T telephone system is provided for communication within and outside any organization or operator company. Incase of failures of internal communications satellite phone can be used.

4.6 DESALINATION AND POTABLE WATER SYSTEM MODULE

Potable water is required for personnel on board and for generating steam. Accordingly a potable water system is provided at FPS which comprises of water maker, filters, pressurized vessels and storage tanks. The sources of potable water for these floating production system can be obtained either by desalination of the seawater or by transportation of fresh water from land. However, water received by transportation from land vary in quality and quantity, and most important of all, can be very costly. So desalination of the sea water is the best option as source water.

Seawater must be desalted before it can be used for most process applications and for drinking purpose. There are six common processes for desalting seawater, the most common being "Reverse Osmosis" process. These are: (a) Mechanical vapor compression (b) Thermo-compression (c) Multiple effect (d) Once-through, multistage flash with brine recirculation and (e) Reverse osmosis process.

The most commonly used process for desalting seawater is the reverse osmosis process, based on membrane technology. RO process is a membrane process for removing 95–99% of all dissolved minerals, 95–97% of dissolved organic materials and 98% of biological and colloidal matter. The flow is reversed by applying pressure to the seawater (feed side), which is more than the osmotic pressure. Only pure water flows through the membrane. Reverse Osmosis (RO) type units, using permeate especially for seawater has the advantage of reduced maintenance and high reliability. Also, their operation is not affected by the motion of the vessels. RO units are also priced competitively compared to other types of desalination units. This process yields little or no pollution problems, as the discharge contains only 30% more salt than the original seawater. Further, Boiler feed-water is also drawn from this unit. The steam production will largely depend on the system chosen and on the system available.

Normally the Reverse Osmosis (RO) type water makers consist of the following:

- Sand filters
- Cartridge filters
- Chemical feed
- High-pressure pump
- Permeator

The RO units separate suspended and dissolved solids from raw sea water and make it potable. The sand filter or the diatomaceous earth filter is used to remove the suspended solids. The cartridge filter is an additional protective measure to ensure filtration of the particles, if any has passed through the DE filter. The high-pressure pumps are utilized to increase the seawater pressure in the permeater. The RO module or permeater is a high-pressure fiber glass vessel which houses the polymeric material that acts as the membrane. The feed water prior to filtration is dosed with chemicals like H_2SO_4 or PH adjustment and coagulant for agglomerating the suspended mater sale. Thereafter the solids are filtered out in the DE filter and are further polished by the cartridge filter. The high-pressure pump increases the pressure of the feed water for passing it though the RO module. The potable water thus produced is chlorinated before storing in the potable water storage tanks. As the membranes are highly susceptible to chlorine and the raw seawater contains chlorine hence $NaHSO_3$ is also injected to remove free chlorine from the seawater. In addition anti-scalant is also dosed to prevent $CaCO_3$ and Mg_2CO_3 deposition on the membrane. The product water (potable water) should contain less than 500 ppm of TDS. Hypochlorite is generated by electrolysis from seawater feed and is then distributed to the various users of potable water system, where it helps to prevent micro-organic growth.

The water consumption per man on-board FPS is estimated to be about 40 US gallons per day. Accordingly, total requirement is worked out for the entire crew and accommodation.

4.7 UTILITY WATER SYSTEM MODULE

Utility water pump which normally is a submersible electrical centrifugal pump for lifting sea water and wash down pump that normally acts as standby to utility pump, are used for feeding water to:

- Water maker
- Chlorinator
- Wash down hose reel
- Living quarters
- Sewage treatment
- Firewater header

Normally the requirement is met by the utility water pump, but in case of heavy drawl, the pressure in the utility will drop and a pressure switch (PSL) actuates which automatically starts the wash down pump. Chlorinator generates NaOCL (hypochlorite) from seawater by electrolysis. Thereafter NaOCL is injected at the intake of wash down, utility and firewater pumps to take care of bacterial growth in the system. In case both the wash down and utility pumps stop, then water maker and chlorinator will trip. With the stoppage of these pumps the firewater header pressure will drop gradually and if this decreases below certain predetermined value the selected firewater pump will start automatically.

4.8 COOLING WATER SYSTEM

Potable water from the potable water storage tank is used to cool different pumps like Hot oil pumps, Glycol booster pumps, Glycol recirculation pumps etc. The system comprises of a cooling water tank, cooling water pump and cooler. The cooling water tank is an atmospheric vessel and the makeup is automatically done through a float type level control valve. The cooler is of finned type and is cooled by fan driven by electric motor.

A separate cooling system is put in place on board as it is not recommended to use seawater directly as a cooling medium, since the seawater is highly corrosive. The cooling medium should be fresh water, with or without the addition of glycol and corrosion inhibitor. This cooling medium will be cooled by a seawater heat exchanger made of corrosion-resistant.

4.9 SEAWATER SYSTEM

Seawater is used to produce potable water, to cool the inert gas generator, to generate hypochlorite, to pressurize the firewater main, and is also used as service water.

4.10 CHEMICAL INJECTION SYSTEM MODULE

PPD (pour point depressant), demulsifier, oil and gas corrosion inhibitors are injected into the crude oil and gas streams. Oxygen scavenger and bactericide injection systems are also used.

4.10.1 PPD (Pour Point Depressant)

If crude is of high pour point nature, pour-point depressant gets injected in the crude stream if heating is not available due to equipment failure or during planned shutdown of production. This unit facilitates start-up of facilities after a prolonged shutdown.

A system is provided for injection of a PPD agent (for example, say Shellsivim-5 X) into the suction header of the booster pumps prior to main oil line pumps. Injection system is designed for a maximum dosing rate say of 500 ppm based on a crude flow of, say 50,000 barrel per day for two booster pumps. PPD is stored in a PPD storage tanks having a pre-determined minimum storage capacity. PPD storage tanks are provided with a heating coil using hot oil as heating medium and mixer. The PPD piping is maintained at constant temp with electrical heat tracing.

4.10.2 Demulsifier

A system is provided for injection of demulsifier (for example, say, Diatrolite DE-220) into oil manifold. The injection facility is designed for a maximum dosing rate say, of 300 ppm based on the maximum incoming fluid flow to the surge tanks. The demulsifier storage tank is provided with an electrical heater and a mixer capable of maintaining a temperature of 60°C.

4.10.3 Corrosion Inhibitor

Depending upon the ppm of hydrogen sulfide, percentage of CO_2 and water in the well fluid, the corrosion inhibitor system is designed and put in place. It also takes into account formation water that is likely to be produced with the reservoir fluid in the later life of the field. A batch treatment of corrosion inhibitor is injected into the well tubing periodically to prevent corrosion from saline formation water or hydrogen sulfide. Batch treatment of corrosion inhibitor gets accomplished by back-flowing the well through the flow line, with either dead crude oil or diesel as corrosion inhibitor diluents. Adequate pumping of diesel or crude should be readily available from the tanker utilities; however, it requires a proper tie-in connection. A hydrocarbon-soluble corrosion inhibitor may be mixed in a 1–10% mixture, with diluents to be injected in the flow line and well tubing. Facilities required for corrosion inhibitor consist of one storage tank, corrosion inhibitor injection pump, mixer, and various instrumentation.

Corrosion inhibitors are of two types: Oil Corrosion inhibitors and Gas Corrosion inhibitors:

* *Oil Corrosion Inhibitor:* A system is provided for an injection of oil corrosion inhibitor (for example, say COREXIT 7730) into the downstream of production manifold and into the oil stream downstream of surge tanks, at a designed dosing rate, say of 60 ppm or so.
* *Gas Corrosion Inhibitor:* A system is provided for injection of gas corrosion inhibitor (for example, say COREXIT 7730) upstream of each second stage gas cooler and each third stage gas cooler of the gas compression trains. Injection is also provided down stream of each glycol contactor overhead scrubbers, at a designed dosing rate say, of 16.7 L/ MM $Nm^{3\,or}$ or so.

Facilities are also provided for the injection of some other chemicals as planned, desired and required. Some of them are Oxygen Scavenger, Bactericide, and Chlorine etc.

4.11 CRUDE OIL WASHING SYSTEM (COW)

For the tanker-based FPSO system, a COW system, complying with the requirements of IMO Tanker Safety and Pollution Prevention (TSPP 1978), is used for the sludge control and for cleaning of cargo tanks.

The COW system is arranged such that it can be supplied with crude oil by any one of the cargo pumps and has facilities for connecting into the tank-cleaning heater. The system serve as fixed deck-mounted and submerged tank-cleaning machines in all cargo tanks, including the slop tanks. Each tank-cleaning machine is to have its own isolating valve. An additional stripping eductor is fitted and the piping are arranged to enable the eductor to be driven by any of the cargo pumps and to discharge into the slop tanks and one of the aft cargo tanks. Instrumentation control is there to enable the crude oil washing operation to be monitored from both the cargo control room and the main deck at the top of the cargo pump room. An access platform or ladder is to be provided to all submerged tank-cleaning machines which are not readily gets decided as per the requirement of the vessel registry administration.

4.12 PURGE GAS SYSTEM

To ensure a safe explosive level inside the storage tanks, it is necessary to prevent air from being drawn into the storage space. This can be achieved by having a good purging system in place and is achieved by connecting the vacuum side of the pressure/vacuum relief valve (breather valve) to the inert gas system. Each cargo tank is to be fitted with a pressure/vacuum relief valve, or a vent pipe is led from each tank into a common header. In the latter case, the header is to be led to a reasonable height above the deck and is to be fitted with a flame arrestor and a pressure/vacuum relief valve at the outlet to the atmosphere.

During production at designed rates, the vapor displaced from storage should be sufficient to prevent air ingress. However, during tanker loading and periods of low production, it will be necessary to purge the storage tanks to maintain the minimum required forward flow. It follows that during normal production the purge gas system could be shutdown (or operating at minimum turndown). However, during shuttle tanker loadings, it should be capable of supplying in excess of the equivalent loading rate to the storage system.

This purge gas supply should ideally be oxygen free and can be either hydrocarbon gas or inert gas. During production shutdowns, hydrocarbon gas will not be available. For this reason and with the desirability of minimizing the hazards of hydrocarbon gas leakage, an inert gas purge system is always preferred. The most appropriate method of generating inert gas for this facility is through the use of combustion-type inert gas generators. These units generally operate on the principle of the stoichiometric combustion of fuel to produce gas containing less than 0.5% oxygen. The gas is generally cooled by direct contact with water.

4.13 VAPOR RETURN SYSTEM MODULE

As an alternative to a purge gas system, a vapor return system, plus a small inert gas system, is considered so as to have a safe explosive limits inside storage of a tanker. The advantages of such a system over a purge gas system is that it has lower investment cost because of a much smaller inert gas generator requirement and also because of elimination of emission of gases into the atmosphere during tanker loading.

A vapor return system consists of a vapor line running back to the storage tank from the tanker. During tanker loading, the (inert) gases from the shuttle tanker are recycled back to the storage tanks at the same rate as the crude is pumped to the tanker. During production, the gases (that came originally from the tanker) are gradually displaced and vented to the atmosphere through a breather valve at a slightly elevated pressure.

Only temperature variations of the storage tank are likely to cause some in-breathing of air. Therefore, when the storage tank vapor pressure is below atmospheric pressure, inert gas is required in the storage tanks to prevent air ingress. The inert gas for this purpose can be supplied either by nitrogen bottles or from a small inert gas generator. It may be advantageous to combine the inert gas systems with a vapor return system for economic reasons.

4.14 DIESEL FUEL SYSTEM

Diesel or gas oil liquid fuel is an alternate fuel source for main power generation and stripping pump fuel when fuel gas is not available, as happens during shutdowns and start-up. Diesel fuel is assumed to be stored in a storage tank on the FPS for a minimum period, equal to or exceeding the interval between arrival of the shuttle tankers. The required diesel consumption and storage capacity are shown below. This is assuming a 7 day shutdown situation during which diesel is required for emergency power, heating the cargo, and offloading of crude.

- FPSO Barge Concept
  - Approximately 135 tones of diesel fuel are required.
  - Approximately 25 tones of diesel fuel are required for flow line flushing
- FPSO Tanker Concept
  - Approximately 170 tones of diesel fuel are required
  - Approximately 55 tones of diesel fuel are required for flow line flushing
- Semi-submersible FPS Concept
  - Approximately 170 tones of diesel fuel are required
  - Approximately 15 tones of diesel fuel are required for flow line flushing

Recommended storage capacity for barge FPS   :   175 tones
Recommended storage capacity for tanker
and semi-submersible FPS's   :   250 tones

Figures here are only indicative just to develop an understanding.

4.15 BALLAST WATER TREATMENT SYSTEM MODULE

The water from the shuttle vessels must be treated at the FPSO's loading facility. If the shuttle vessels are equipped with a segregated ballast system, then the ballast water can be dumped overboard while offloading. Otherwise, ballast water may be offloaded from the shuttle vessel and loaded onto the FPSO vessel. Produced water tank compartments are to be used to accommodate the ballast water from the shuttle vessels. Oil removed from these compartments will be recycled back to the main separation train.

The International Marine Consultative Organization (IMCO) has proposed that a maximum oil content from tanker deballast water be at 15 ppm, or less. This is less than, or equal to, "oil sheen" water quality, depending on temperature. The 1974 International Convention for the Prevention of Pollution of the Seas by Oil prohibits the discharge of ballast water containing more than 100 mg per liter of oil restricted coastal areas.

4.16 HOT OIL SYSTEM MODULE

The hot oil is an organic liquid with very high boiling temperature and is used as an indirect heating media.

The hot oil system comprises of waste heat changer, hot oil expansion tank and pump. The expansion tank is a closed atmospheric vessel, wherein hot oil is filled to charge the system. Hot oil temperature goes up to 300°C and it is to be ensured that it should not come into contact with air; therefore the expansion tank has a nitrogen gas blanket.

The hot oil from the tank is pumped to the Waste Heat Exchanger (WHE). The WHE are large exchangers. The flue gas from the gas generator is utilized to heat the oil. The flue gas can be released to the atmosphere directly or through the waste heat exchangers. The temperature switches and SDVs take care that is should not get overheated. Even if the diverter valve towards the WHE is closed still there is possibility that the flue gas may leak through the WHE, hence the valves are so designed that always a minimum flow of hot oil should be through the WHE. Excessive heating can lead to cake formation. The FSL and TCV take care of this aspect. The temperature of the hot oil after WHE is high and goes to the glycol re-boiler, this header can be termed as High Temperature and the one going to water trough and chemical tank can be termed as low temperature header. Thereafter the hot oil goes back to the WHE and this header can be termed as return header.

The PCVs and TCVs in the system, as well as the flow switches takes care that required heat is available to all the units and a minimum flow across the WHE is always maintained. The individual units are also equipped with SDVs, which on actuation of the respective temperature switch gets closed.

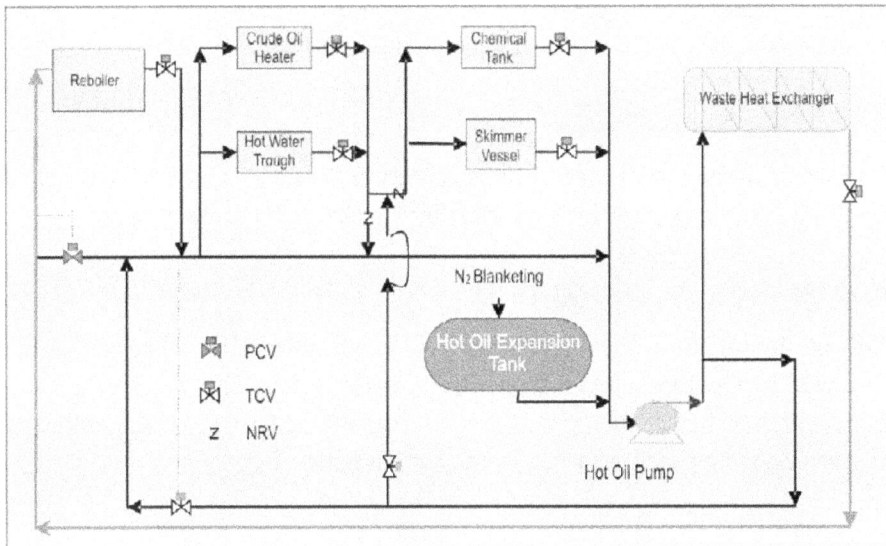

Hot Oil System

Generally the following units on any FPS are required to be heated.

4.16.1 Glycol Re-Boiler

The rich glycol is to be stripped of water before it is again circulated. The glycol re-boiler heats the rich glycol and with the stripping column and stripping gas the water vapor goes out through the glycol still and the regenerated lean glycol is stored in the tank below the re-boiler. This is the major consumer of hot oil. Therefore the high temperature header takes care of heating the glycol re-boilers. The re-boiler temperature is maintained around 204°C.

4.16.2 Crude Oil Heater

The crude coming out of the separators are required to be heated to aid in the demulsification process in the order of priority, this is the second in the consumption of heat from hot oil. The temperature is around 60°C.

4.16.3 Skimmer Vessel

The crude from the closed drain header and the sump caisson is sent to the skimmer vessel. There is every likely that this crude can congeal or become semisolid. In that case taking the liquid out of the skimmer will be difficult. The temperature is therefore maintained around 50 °C with the help of hot oil.

4.16.4 Chemical Tank

Only PPD, sometimes gets congealed if a temperature of about 50°C is not maintained. The consumption of heat is comparatively less in this case.

4.16.5 Hot Water Through

The water in the hot water trough is heated to about 60°C–65°C in order to melt the PPD in the barrels and transfer it to the PPD storage tanks.

4.17 SEWAGE PLANT MODULE

Sewage treatment/effluent treatment plants are installed to chemically and physically treat the sewage from the living quarters and the waste from galley/kitchen before being discharged into the sea to meet the statutory environmental regulations for offshore installations.

The living quarters is usually serviced by a sewage treatment plant. For a stationary vessel, sewage treatment is considered essential to avoid contamination of inlet seawater used for the desalination unit(s). Three types of sewage plants are in practice. A conventional sewage treatment plant treats all the waste water. A vacuum type or chlorine-contact type treats only the "black" water from toilets or sick bay. "Gray" water from showers, basins, and galley use is dumped directly overboard. The use of a conventional system is normally preferred because it is the most reliable system and provides the least pollution in all three FPS concepts.

4.18 KILL FACILITIES

During the initial two or three years of production, adequate wellhead flowing pressure may create a hazard if a leak develop between the tubing and casing. To contain this pressure and put the well in a harmless state for work-over, a well-kill facility is being provided on FPS. The pressure required for well killing is dependent on the particular situation. Well-kill fluid is filtrated seawater. Weighted drilling mud is not required for the relatively shallow reservoir.

4.19 HELIDECK MODULE

At the FPS, a helideck is installed to allow rapid transport of personnel and material critical to the production facilities. The helideck must be capable of handling the state-of-art helicopter say for e.g. a Silkorsky-61 N/L helicopter and accordingly dimensions of helideck are determined. For example, for, Silkorsky-61 N/L helicopter the size of the helideck should be 25 m × 25 m (73 ft × 73 ft) and should weigh approximately 240 kips (109 tons). The landing area is to be readily accessible from the accommodation, and the take-off and approach path is to extend over an area of 210 degrees. This area must be clear of obstructions. Two separate means of access to the landing area are to be provided. Lighting is to be provided to allow night operations. The landing area should be delineated by alternate yellow and a safety net is to be provided at the edges. A rope net with suitable anchorage points is to be provided on the upper surface.

Re-fuelling facilities which are compatible with the operating procedures for a given helicopter are also provided and recognized fuel quality control procedures have to be adhered to. The facility generally consists of a dispenser unit, a transportable tank and facilities for handling the tank, as required. Firefighting facilities has to be provided in accordance with ABS Rules for Industrial Process Plants. Foam monitors are to be provided. Fuel storage areas are to be protected by a fixed fire detection and protection system.

4.20 ATF REFUELING SYSTEM

The ATF refueling system comprises of storage tanks, pump, filters, meter and hose reel. Normally the tanks and pumps are located in the lower or middle deck. At the time of refueling the helicopters, the pump can be started from the helideck through a push button switch. The ATF from the pump passes through a filter and a meter located near the helideck before it enters the nose reel. The nozzle of the hose reel is required to be earthed in order to avoid any sparking, which can be a source of fire at the time of refueling.

4.21 MATERIALS HANDLING SYSTEM

The following equipments are normally available on any FPS for the purpose of moving material and equipment and assisting in maintenance:
- Pedestal crane
- Electric monorail hoists
- Manual hoists
- Manual trolley hoists

Barge crane

The cranes are also used to transfer personnel, equipment and supplies between barges or supply boats and FPS.

4.22 HEATING, VENTILATION AND AIR CONDITIONING EQUIPMENT MODULE

For efficient functioning of control systems installed for various equipment in process control rooms and turbine generator control rooms and for safety reasons to avoid ingress of inflammable hydrocarbons in case of emergency, all closed and living quarters are kept pressurized and a temperature of around 25 degree centigrade is maintained through Heating Ventilation And Air Conditioning (HVAC) equipment like air handling units, chiller handling units and pressurizing units. In addition, all the living quarter areas like residential rooms, recreation rooms, galleys, offices etc. are also temperature, humidity and pressure controlled.

4.23 DRAIN SYSTEM

There are three types of drains:

- Deck drain
- Closed drain
- Condensate drain

The deck drains are open drains, which are connected through a header. Water is used for cleaning decks. This dirty water and other spillages on the decks are dumped into the sump caisson through the deck drain header. The closed drain header is a low-pressure header and the liquids thus drained also go to the sump caisson/shimmer vessel. The condensate drain header is a high-pressure drain header. Liquids from high-pressure vessels like dehydration system, pressure gas compressor, KOD (knock out drums) etc, are drained through this header to the surge tank.

Floating Production System: Safety and Fire Fighting System/Facilities

Floating Production System are covered with various safety systems and the concept of the total system is based on quick detection of unsafe situation followed by prompt remedial actions so as to bring the situation under control and resumption of normal operation. The system and equipment on FPS are designed for safe and reliable operations. All personnel concerned with the operations and maintenance should be suitably qualified and thoroughly familiar with the details and operating characteristics of the systems and equipment.

In designing any FPS, the primary consideration is given to safety of personnel, environment and facilities. The release of hydrocarbon is a factor in virtually all threats to safety. Thus, the major objective of the safety system should be to prevent the release of hydrocarbon from the process and to minimize the adverse effects of such releases, if they occur. The design is based on API-RP-14C. The main objectives of any safety system in place are:

- To prevent any undesirable event that could lead to release of hydrocarbon.
- Shut-in hydrocarbons to a leak or overflow, if it occurs.
- Accumulate and recover hydrocarbon liquids and disperse gases that escape from process.
- Prevent ignition of released hydrocarbons.
- Shut-in the process in the event of fire.
- Prevent undesirable event that could cause release of hydrocarbons from equipments other than that in which the event occurs.

Before we proceed further, let's appraise ourselves with some basic definitions:

(i) *Safety:* Control of accidental loss or hazards. It involves constant awareness to critical work hazards through a constant improving system.

(ii) *Hazard:* The potential to cause harm to people, property or environment.

(iii) *Hazard Effect:* Any loss of life, profits, business, reputation, skills etc.

(iv) *Risk:* Risk is a combination of the Hazard effect and the Probability that Harm to People, Property or Environment will actually occur.

Risk = Hazard effect × Probability of occurrence

(v) *Accidents:* Undesirable events that results in harm to human being, damage to property or environment.

SAFETY AND FIREFIGHTING FACILITIES/SYSTEMS AT FLOATING PRODUCTION SYSTEM

Let's elaborate over the Safety and Firefighting systems that have to be on any floating production systems, either in all comprehensiveness or in some exclusivity depending upon the process, systems and field requirement:

5.1 Safety Systems

    5.1.1 System and Equipment Safety System

    5.1.2 Emergency Shutdown (ESD) and Fire Shutdown (FSD) System

    5.1.3 Personnel Safety System and PPE (Personal Protective Equipment)

5.2 Fire and Firefighting System

    5.2.1 Fire Fighting Systems/Principles

    5.2.2 Fire Detection System

    5.2.3 Fire Suppression System

5.3 Hydrogen Sulphide Safety

5.4 Emergency Plans

Before we discuss each of these in detail, I would like to deliberate upon one very important concept here, i.e., the concept of "work permit system", that goes a long way as far as safety of platform and floating production systems and vessels are concerned.

Work Permit System

The work permit system is an important tool for safety in hydrocarbon processing/handling. If any work has to be performed in a hydrocarbon processing installation by any person other than the operating personnel of that area, a duly, authorised written permit shall be obtained by the person/agency executing the work before commencement of the work. The work permit system contain the information like job description, location of work, compliance of requirement for location safety and personnel safety, emergency plans and responsibilities, start and end time of the work and similar other information, duly signed by the contractor or executor of the work and the in-charge safety of floating production system.

Work permits are of two types: Hot work permit and Cold work permit

Hot Work: Hot work is an activity, which may produce enough heat to ignite a flammable air-hydrocarbon mixture or a flammable substance.

Cold Work: Cold work is an activity, which does not produce sufficient heat to ignite a flammable air-hydrocarbon mixture or a flammable substance.

5.1 SAFETY SYSTEM

5.1.1 System and Equipment Safety

The various systems and equipments are fitted with different safety devices like:

- Well and well head   : SSSV and SSV
- Flow line   : PSH, PSL, PSV
- Headers   : PSH, PSL, PSV
- Pressure vessels   : PSH, PSL, PSV, TSH, LSH, LSL,ILSL
- Fired vessels   : PSH, PSL, PSV, TSH, LSH, LSL
- Pump   : PSH, PSL, PSV, SDV
- Compressor   : PSH, PSL, PSV, SDV, BDV, TSH, LSH, LSHH, LSL, LSLL

The abbreviated form carries its usual meaning of instrumentations for oil and gas industry; like PSL: Pressure Switch Low; PSH: Pressure Switch High; SDV: Shutdown Valve etc.

SAFE Chart

Let's understand through the table given on next page how safety analyses are undertaken with respect to equipment/facilities/systems. A Safety Analysis Function Evaluation (SAFE) chart is made for the given FPS, which related all the equipments and the sensing devices, shut down devices and emergency support system to their function. The system in general provides two levels of protection i.e. primary and secondary to prevent or to minimize the effects of an equipment failure with in the process.

5.1.2 Emergency Shut Down System and Fire Shutdown System

Emergency Shut Down System

An ESD is a system of manual control located on a FPS which when actuated will initiate shut down of all wells and other process systems. ESD system provides a means for personnel to initiate process shut down of a FPS when an abnormal condition is detected. In case of actuation of ESD all process operations will stop, SDVs, MOVs Control valves will go to the fail-safe position, BDVs will open and vessels will get depressurized. The emergency shut down system consists of a pneumatic loop, kept pressurized at a given critical pressure specific to the chosen FPS, say for example, 40–50 psig and goes all around the FPS. When actuated, it initiates shut down

Safety Analysis Tables

Undesirable Event	Cause	Detectable Condition at Component	Protection Primary	Secondary
(A) *Pressure Vessels:*				
Over pressure	→ Inflow exceeds out flow → Thermal expansion → Blocked outlet	High pressure	PSH	PSV
Under pressure	→ Withdrawals exceed inflow → Thermal contraction	Low pressure	Gas make up system	PSL
Over flow	→ Liquid inflow exceeds liquid output capacity → Level control failure	High pressure High level	LSH	*LSH and *PSH
Gas blow by	→ Level control failure	Low liquid	LSL	*PSH and *PSV or events
Leak	→ Deterioration → Rupture accident	Low pressure and backflow	PSL	ESS
Excess temp. (process)	→ Excess heat input	High temp. (process)	TSH	Safety devices on heat source
(B) *Atmospheric Vessels:*				
Over pressure	→ Blocked or restricted outlet → Inflow exceeds outflow → Gas blow by (upstream component) → Pressure control system failure → Thermal expansion → Excess heat input	High pressure	PSH	PSV/PRV
(C) *Pumps:*				
Over pressure	→ Blocked discharge line	High pressure	PSH	PSV
Leak	→ Deterioration rupture accident	Low pressure and backflow	PSL and FSV	ESS
(D) *Compressors:*				
Over pressure (suction)	→ Failure of upstream pressure control device	High pressure	PSH (suction)	PSV
Leak (suction)	→ Deterioration → Rupture accident	Low pressure High gas concentration	PSL (suction) ASH (building)	ESS
Over pressure	→ Blocked discharge line excess back	High pressure	PSH (discharge)	PSV (discharge)
pressure Leak (discharge)	→ Deterioration → Rupture accident	Low pressure and backflow high gas concentration (building)	PSL (disch.) and FSV (disch.) ASH (building)	ESS
(E) *Pipelines:*				
Over pressure	→ Blocked discharge line	High pressure	PSH	PSV
Leak	→ Deterioration rupture accident	Low pressure and backflow	LSL and FSV	ESS

*ESS—Emergency Support System

of the complete facility. This ESD either needs to be actuated manually or should get actuated automatically if emergency so arises or if any vulnerable process upsets might get witnessed. Let's understand here some of the ESD actuation options. It can be actuated by:

(a) Actuation of 'pull of ESD' buttons, which are located in all the strategic locations of the FPS and also in the control room. When actuated, pressure in the loop decreases and actuated a pressure switch in the main shut down panel.

A typical Emergency Shut Down (ESD) station

(b) It can also be actuated electrically by a solenoid valve from the control room. Thereafter the main shut down panel sends signal to the individual control of panels of the various components and the emergency shut down takes place.

(c) Besides these two actuation, there are some process situations where actuation takes place automatically:

- High level in flare KOD, otherwise oil will go the flare
- Signal from fire and gas panel, when LEL is more than 60% or 50 ppm of H_2S
- Actuation of PSHL of main line

(d) In the following emergency conditions, the ESD/FSD system should be actuated by any person on board:

- Fire on a platform
- Leakage in main oil/gas line
- Blow out
- Leakage of sour gas, H_2S concentration of 20 PPM or above in environment

(e) However, there are some other conditions too, when ESD/FSD needs to be actuated but only after the permission from Oil Installation Manager or Field Production Superintendent (highest authority at FPSO):

- Cyclone and severe weather conditions
- Fire in a nearby installation
- Oil spill around the installation
- Collision involving the installation
- Rupture of pipe/uncontrolled oil or gas leakage

Fire Shut Down System (FSD)

Similar to ESD system, a Fire Shut Down (FSD) system is also provided at the FPS. It is a system of manual control, in addition to control from various fire sensing devices. This system consists of a pneumatic loop running through out the FPS. The loop is normally kept pressurized and comprises of fusible plugs. In case of fire the fusible plug melts resulting in loss of air pressure in the loop and actuates the FSD system. In addition to this and ESD pull buttons, FSD pull buttons also exist on the platform at strategic location, which when actuated leads to FSD.

The following situations lead to fire shut down on the Floating Production System:

- Actuation of UV detector
- Actuation of thermal detector
- Actuation of smoke detector
- Loss of air pressure in FSD loop by melting of fusible plug

We will be discussing these detectors in some details in this chapter ahead.

In case of fire shut down, total facilities (process and utilities) on the FPS comes to halt. The sprinkler system actuates and thus the pressure of firewater header reduces which in turn starts firewater pump. Power for the pump in this situation is being supplied by emergency generator.

ESD/FSD Stations

The ESD/FSD stations should be conveniently located but should be protected against accidental activation. It should be clearly identified. The stations are generally located at helidecks, exit stairways, boat landings, muster stations, near the main exit of living quarters. The ESD stations should be located on all the decks and at spider deck (at point above the monsoon storm damage zone). The paging/communication system of floating production system is utilized for all emergency alarms to ensure that personnel on board are aware of emergency conditions that might arise.

5.1.3 Personnel Safety Systems

Safety of personnel is the most important factor in the operation of any industry. General safety rules are practiced and enforced for all personnel on board the FPS as summarized below:

- All personnel in the open deck area shall wear safety helmet and shoes.
- Each person on board shall know where the safety and fire suppression equipments are located and how to operate.
- Smoking shall be permitted only in the specified area.
- Safety belt shall be properly and firmly tied up while working at higher elevation.
- Helideck rules shall be pasted at each entrance to the helideck and should be rigidly followed.
- Escape routes to be prominently displayed at each strategic location.

 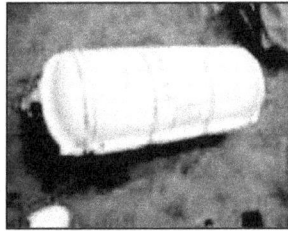 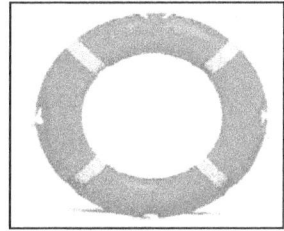

Life boat   Life raft   Life buoy

  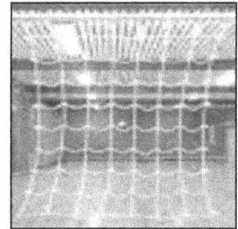

Life jacket   Personnel basket   Scramble nets

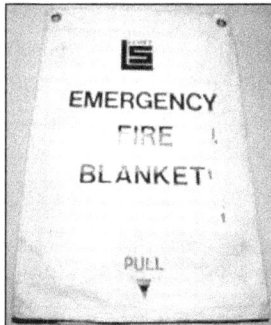 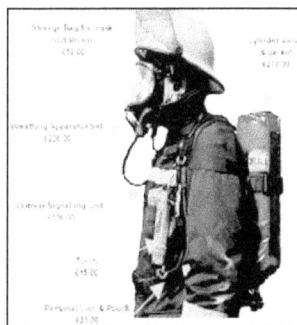 

Fire blanket   Breathing air apparatus   Fire suits

Here, I find it pertinent to mention two **"golden rules"** for operating and maintenance personnel that need to be adhered to for ensuring safe operations and maintenance of system, persons and equipment.

- *Operation personnel golden rule:* Do not open or close any valve without first determining the effect on the field process system.
- *Maintenance personnel golden rule:* Treat each piece of equipment or piping as if it is under pressure and take precautionary measures first.

All floating production systems are provided with adequate life saving equipments and be maintained, tested and kept ready for instantaneous use. The purpose of life saving equipment is to provide safe means of survival in emergency situation in offshore. Towards this end, the following items of life saving equipment are normally available at the floating production systems:

- *Life Boats:* For emergency evacuation from the FPS, fire retardant self-righting lifeboats are provided on all the FPS. There location is supposed to be known to each individual on board. In case of any emergency the instruction of boat supervisor/captain is of most importance and must be followed without any delay/hesitation.
- *Life Raft:* Inflatable life rafts are provided on FPS as a back up of lifeboats. Life rafts are positioned strategically and are easy to operate. The life rafts are inflated automatically when they are thrown into the sea by an automatic mechanism. People on board then jump to sea wearing life jackets, swim towards the raft and then load themselves into the raft.
- *Life Buoys:* These are round floatation rings fitted with automatic light. These rings are used for life saving when a man accidentally goes overboard. Person on board throws these buoys to the sea near the drowning person and the drowning person put all efforts to catch hold of it and keep holding till the time they are picked up by the rescue boat.
- *Life Jackets:* These are floatation jackets, which are worn for floatation. These life jackets are supposed to be worn whenever working on the edges of floating production systems or whenever emergency arises.
- *Personnel Baskets:* These are specially designed and certified baskets for personnel transfer from boat to platform or boat to boat using crane.
- *Scramble Nets:* These are specially designed nets placed strategically on various decks and used for climbing down in an emergency.
- *Fire Blankets:* These are used by personnel on board for escaping a blaze.
- *Breathing Air Apparatus:* Breathing apparatus is used to escape smoke or H_2S emergencies.
- *Fire Suits:* These are specially designed suits used for fire fighting.

The operating instructions and the maintenance procedures for all these safety items are always attached with these items. All persons on board must be familiar with such equipment and location should be easily accessible and prominent.

Personnel Protective Equipment (PPE)

All persons on board FPS must use proper personnel protective equipment while working. Specific PPE is also required to be worn for hazardous operations. Following are the commonly used PPE available at Floating Production Systems:

- Cotton dungarees
- Safety helmets
- Safety shoes
- Hand gloves
- Ear muffs/plugs
- Goggles

Safety helmet

Safety shoe

Ear muffs/plugs

Goggles

5.2 FIRE AND FIREFIGHTING SYSTEM

These systems are designed to ensure the early detection of fires, by way of automatic or manual methods, and of giving early and effective alarm. The fire detection and protection systems are engineered as per the approved codes of practice and should be in accordance with the ABS Guide for Building and Classing Industrial Systems, and SOLAS 74—Safety of Life at Sea 1974.

All enclosed accommodation, machinery and storage areas of the FPSO or FPS will be equipped with fire detector of a type and in numbers suitable for the risk area. Detectors for use in Divs. I and II areas will be either flameproof or intrinsically safe to approved standards. Generally, smoke detectors is employed in areas involving Class 'A' or electrical risks, i.e. accommodation, control rooms, switchgear rooms, and certain machinery spaces. Heat detection takes two forms: rate of rise and fixed. Rate-of-rise detectors are used in general machinery spaces and in accommodation areas of relatively high humidity, i.e. galley and laundry. Fixed heat detection is used in machinery spaces, in areas of high ambient temperature. In addition to the automatic fire detectors, manual alarm contacts (break-glass units) is located in all areas, generally at points of access or egress. All detection zones are hard-wired to a fire control panel located in the central control room. The fire panel will give both audible and visual indication of the location of any alarm and will be complete with the desired logic-giving alarm outputs and equipment and/or ventilation shutdowns. The panel will also indicate system faults.

It's important to understand some basics involved here like that of fire, fire triangle, hazardous area classifications and so on. Refer Annexure C and Annexure D to develop some basic understanding.

5.2.1 Methods of Fire Fighting

Fire can be controlled in any of the following ways:

- *Cooling method*—where fire is controlled by "removal of heat from the surface of fire".
- *Starvation method*—where fire is controlled by "isolation of fuel from fire".
- *Smothering method*—where fire is controlled by "cutting off air/oxygen from surface of fire".

5.2.2 Fire Detection Systems

Fire detection is accomplished through:

- Ultraviolet (UV) detection system
- Thermal and smoke detection system
- Gas detection system
- Fusible plug loops

UV Detection System

UV detectors are utilized to give the earliest possible detection of rapidly developing gas and oil fires. Detectors are sensitive to radiation over the range of 1850 to 2450 angstrom and insensitive to light. The UV detectors are not affected by wind, rain, humidity or extremes of temperature or pressure and are suitable for both indoors and outdoors. They are housed in explosion proof stainless steel enclosures. The UV radiation generated by a gasoline fire with one square foot surface area can be detected at distances ranging from about 15 feet to 50 feet. On the UV controller panel, the probable source or type of fault can be displayed and provide a location code number for the information of maintenance personnel. UV detectors are installed as redundant detector so that all equipment can be supervised by at least two detectors to avoid nuisance action of one detector. To ensure that a fire of sufficient magnitude exists before an extinguishing system is energized, the time delay relay system is used. The time delay relay is field adjustable over the range of 0.2 to 12 seconds. When coincident fire signals are received from two detectors, the FPS audible alarm and Central Control panel annunciator sounds is activated along with causing of FSD.

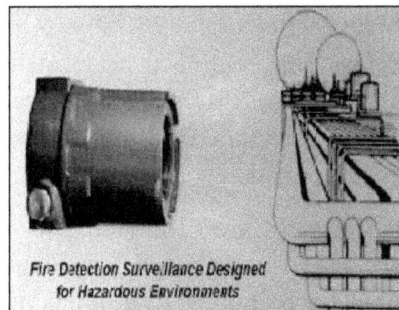

Fire Detection Surveillance Designed for Hazardous Environments

UV detector

Thermal and Smoke Detection System

The thermal detection system detects an increase in temperature caused by a fire, initiates an alarm signal and actuates the automatic extinguishing system for that particular area. When thermal detectors are located in an enclosed room protected with a Halon 1301 system, the signal is transmitted to the fire and gas panel, which in turn produces audible and visible alarms, sends a shutdown signal to the HVAC unit, and actuates the Halon system. The smoke detectors detect both visible and invisible particles of combustion by means of an inner reference and outer sample chamber. The coincident action of two smoke detectors in one protected area actuates the Halon system. Action by only one detector will initiate an alarm only. When detectors operate in room with sprinkler system, audible and visible alarms are activated, but the spray systems operate only when the fire melts the fusible sprinkler head, which then permits water spray on the fire. All detectors are intrinsically safe, explosion proof type.

Thermal detector

Gas Detection System

A gas detection system is provided to detect flammable and combustible gases before they reach a concentration level that would cause a fire or explosion. They are the first line of defense in prevention of human injury and equipment damage. The layout of gas detectors on offshore facilities is determined by two philosophies:

(a) Gas detectors shall be located where the leaks are most likely to occur.

(b) Gas detectors shall also be located where the consequences of a gas accumulation will be greatest.

Gas detector

Gas detectors are calibrated for methane and hydrogen as the expected gases. Gas detectors, which are calibrated, for hydrogen only are installed in the battery rooms. The set points of the alarm are 20% LEL and 60% LEL (lower explosive limit). Alarm is generated on 20% and 60% LEL. Coincident 60% LEL indication by two gas detectors will initiate the ESD of the platform.

Fusible Plug Loops

Firewater header is always maintained at a pre-determined pressure, say for example 10 kg/cm² and all open deck areas in the complex are covered with spray system. Fusible plug loop is ring of pressurized air tubing, which contains fusible plugs at fixed intervals. This loop is covering all the open deck process areas. When these plug comes in contact with heat they melt down and

Fusible plug

pressure of this loop is vented to atmosphere. The fall in the pressure of fusible plug loop activates fire shutdown. Fire shutdown closes down all process systems, generates audiovisual alarms, open the deluge and starts fire water pump.

5.2.3 Fire Suppression and Fire Fighting Systems

Let's understand some of the systems used for fire suppression and for fire fighting.

- Fire water system
- Spray system
- Sprinkler system
- Foam and water hose reel
- Dry chemical fire fighting system
- Halon system
- Fire fighting vessels
- Fire extinguishers

Fire Water System

The fire water system primarily consists of equipment to pump and distribute water for fire fighting purposes. The water for this system is salt water taken from the sea. The pump takes suction from the sea and starts automatically upon the command of FSD from the fire and gas panel. Firewater sprinkler/deluge loops located all over the open areas spray water over the equipment and vessels. Foam water hose reels are

Fire water system

located at various locations throughout the platform so that all areas can be reached from at least two hoses. These hose reels deliver an aqueous film forming foam (AFFF or light water) with seawater to assist in extinguishing spill fires.

The deluge spray protects piping and equipment involved in the processing of oil and gas, as it is important that the equipment be cooled while a fire in the area burns itself cut.

Spray System

Water is the most effective fire fighting medium. The main use is to provide cooling, control the fire and reduce the risk of explosion. The combustible and flammable liquids are cooled to below their flash point and hence the combustion cannot be sustained. A spray deluge system consists of a deluge valve, open type spray nozzles, fusible plugs and fire shutdown valves.

The spray system is actuated automatically or manually. *Automatic Operation takes place by* coincident action of any two UV detectors and by action of fusible plugs. *Manual Operation takes place by* pulling FSD, by pressing the remote manual switch for the deluge valve in the fire and gas panel. Manual operation also takes place in case the automatic deluge valve cannot be operated by the above procedure; the bypass valve is to be opened.

Sprinkler System

The sprinkler systems are installed to extinguish class-A fires, which are likely to occur in the living quarters, workshops and storerooms. The sprinkler system includes manually operated isolation valves, which normally are locked open. If a fire occurs in a room of the living quarters, the smoke detector will be actuated by combustion gas and particles and fusible metal of

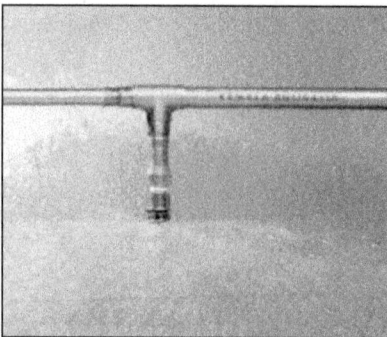

Sprinkler system

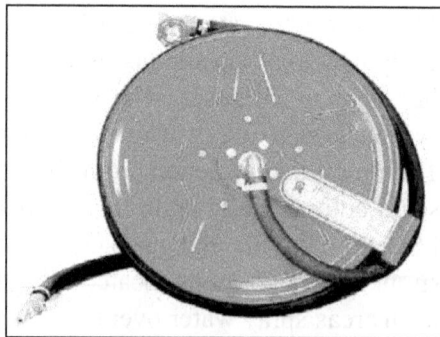

Spray system

the sprinkler head will be melted by the heat generated by fire. Smoke detectors will transmit the signal to the main fire and gas panel. Salt water will be permitted to exit only through the sprinkler nozzles, which have been melted away by the heat, to extinguish a fire. Reduced water pressure will result in the operation of the firewater pump.

Foam and Water Hose Reel System

Water is the cheapest and best medium to cool equipment. A water fog applied to the hydrocarbon surface will flash to steam, giving a cooling and inserting isolation layer effect. If water alone is applied as solid jet stream, most of it sinks below the surface of fuel and how below the surface. This may cause boil over with very dangerous and unexpected consequences of material failure. Aqueous Film Forming Foam (AFFF) is a two dimensional medium, which acts by sealing the static surface of fuel and preventing evolution of flammable vapor.

On the fuel surface, AFFF breaks down rapidly, releasing the film forming solution, which then spreads on the surface of the fuel. AFFF is very effective for use on oil spill fires, but is not effective on gas pressure fires. At least two foam/water hose reels, which consists of an AFFF concentrate storage tank, eductor, hose reel and hose with nozzle, are installed to reach any location on the deck. A foam/water hose reel can discharge either a 6% foam solution or water, only by changing the valves. It can also discharge either for or a straight jet stream by turning the grip handles of the hose nozzle.

Dry Chemical Fire Fighting System

Dry chemical powder is important because it is the most powerful three-dimensional medium available. Powder is effective as an airborne cloud of small particles, which inhibits the chemical reaction of fire. Dry chemical extinguishing systems are considered satisfactory protection for flammable or combustible liquids, combustible gases and electrical hazards. System actuation is possible at each hose reel station or at the storage tank skid by pushing the corresponding actuating device. The dry chemical extinguisher cylinder will be pre-pressurized with a nitrogen back-up supply, which will automatically maintain a constant discharge pressure.

Dry chemical fire fighting system

For operating, reel off the amount of hose length desired, open cylinder valve, direct the chemical at the base of the fire using a sweeping motion to cover the fire area and extinguish the fire.

Halon System

Halon 1301 systems are generally used on floating production system to protect areas where there is an electric fire potential, such as electrical rooms, control rooms, turbine generator enclosures and switchgear rooms, since water is electrically conductive fluid. Halon 1301 agent is a colorless, odorless, nonflammable and electrically non-conductive gas. It is considered non hazardous to personnel when exposed for brief

Halon system

periods at low concentrations (less than 7% by volume). Halon 1301 extinguishers extinguish fire by isolating oxygen from fuel and electrical circuit isolation. The system can be actuated by coincident action of two smoke detectors, single action of thermal detector or by manual activation. Halon is released at a pre-determined interval, say 30 seconds, after operation of the detectors, so that the persons occupying the enclosed areas can evacuate. The fire alarms sound and red light flashes simultaneously, HVAC power is shutdown to prevent the spread of Halon in other rooms and fire dumpers close. As Halon 1301 is not very environmental and equipment friendly, worldwide operators are switching to FM-200.

Fire Fighting Vessels

Big fires are difficult to fight. MSV and Fire Fighting Vessels are having fire-fighting capability for fighting large fires. MSV are dynamically positioned

Fire fighting on a rig

Testing of fire monitor

and normally have 4–5 remote controlled fire fighting pumps with capacity, say for example of 1800 m³/hr with 4 water monitors and 1 foam concentrate pump of capacity, say for example of 300 m³/hr, which is a sufficient capacity to cater to any fire fighting situation in the field. Care is taken to position these vessels near the field during normal inspection/repair jobs, to enable one of these vessels to reach the emergency location within an acceptable intervention time of say one or two hour.

Fire Extinguishers

Portable fire extinguishers are designed for small fires and are used in close proximity of burning materials. Various types of extinguishers are as follows:

(a) *Water Extinguisher*

Plain water expelled by pressure released from CO_2 cartridges. These are useful for class-A type fire.

(b) *Foam Extinguisher*

1. *Chemical Foam Extinguisher:* It consists of inner and outer container with cap assembly, outer container hold sodium bicarbonate ($NaHCO_3$) and inner container contains solution of aluminum sulphate $Al_2(SO_4)_3$. On operation two chemical reagents get mixed up, carbon dioxide is liberated and froths are produced which is expelled by CO_2 through nozzle.
2. *Mechanical Foam Extinguisher:* In this type of extinguisher foam concentrate is stored in a sealed container. When the extinguisher is activated, foam solution is expelled by pressure release into the body of the extinguisher from CO_2 cartridge or by pressure maintained in the

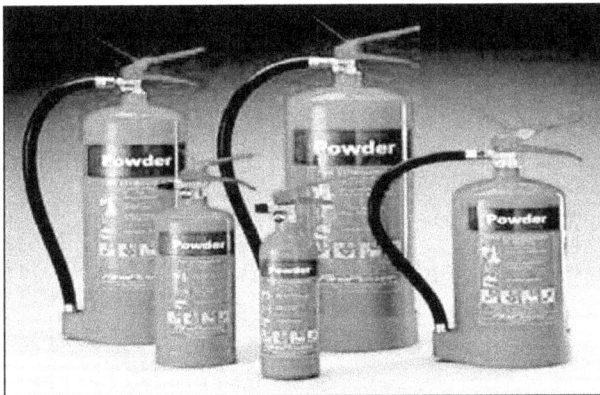

Dry chemical powder extinguishers

body of the extinguisher by air or nitrogen. This type of extinguisher smothers flames with foam blanket and can be used on class-B fire and also on small class-A fire.

(c) Dry Chemical Powder Extinguisher

These are again of two types: (i) The powder is released into the body of the extinguisher from CO_2 cartridge, (ii) The powder is expelled by pressure maintained in the body of the extinguisher by nitrogen (stored pressure). The powder knocks down the fire immediately. The extinguisher is suitable for use on class-B, C and D fires and also on small class-A fire.

(d) Carbon Di-Oxide Extinguisher

The extinguisher consists of carbon dioxide (CO_2) in a pressure cylinder, a tube and valve for releasing the CO_2 and a discharge horn. These extinguishers are intended for use of class-B and class-C fire however can be used on small class-A fires.

(e) Halon Extinguisher

In general liquefied gas fire extinguishers i.e. bromo-trifluro methane (Halon 1301) and bromo-chloro di-fluoro methane (Halon 1211) have features and characteristics similar to CO_2 fire extinguishers. This type of extinguisher in non-corrosive and leaves no residue. On weight basis it is twice effective as compared to CO_2. This type of extinguisher is suitable for use on class-B and C type of fires.

5.3 HYDROGEN SULPHIDE SAFETY

Hydrogen Sulfide (H_2S) is a toxic gas that leads to unconsciousness and to death if inhaled or exposed substantially to a threshold limit. Threshold Limit is that concentration level at which it is believed that all workers may be repeatedly exposed day after day without adverse effects.

The presence of H_2S is a growing concern in the oil production. New fields are being developed and oil fields are being treated in areas where H_2S can be expected to increase probability of exposure. Hence it is

Use of self contained breathing apparatus (SCBA)

important to understand "Hydrogen Sulphide Safety" comprehensively including the hazards of H_2S, its characteristics, how the H_2S are detected, what protections are available if working in H_2S environment and what working standards are there for H_2S environment and what are the locational safety and emergency rescue available. Besides this, we also need to understand how H_2S is affecting the metals and what the materials that suits for H_2S environment are. In subsequent paragraphs, all these have been explained briefly so as to develop some understanding.

5.3.1 H₂S: Hazards and Characteristics

H_2S is extremely toxic gas and rank second only to hydrogen cyanide. The Principal Hazard of H_2S is DEATH by inhalation. When the amount of gas absorbed into the blood stream exceeds that which is readily oxidized, systemic poisoning results with a general action on the nervous system. Labored respiration occurs shortly and respiratory paralysis will follow immediately at higher concentrations. Death will occur from asphyxiation unless the exposed person is removed immediately to fresh air and breathing stimulated by artificial respiration. Other level of exposure may cause the following symptoms individually or in combination: Headache, dizziness, nausea, cough, drowsiness, eye and throat irritation, dryness and pain in nose, throat and chest.

Detection of H_2S solely by smell is highly dangerous as the sense of smell is rapidly paralyzed by the gas.

Refer to Annexure E to understand the H_2S hazards, its characteristics and the physical effects of different H_2S concentration level. Besides its impact on human body, it's also important to understand the effect of H_2S on metals and the material selection pre-requisites for H_2S environment. A brief over this has been given at Annexure F.

5.3.2 H₂S: Detection

Following are some common detection devices

Lead Acetate, Ampoules or Coated Strips

Lead Acetate paper detectors operate under the principle of contamination. When H_2S comes in contact with the lead acetate coating on the paper, lead sulphide is produced and will turn to various shaded from tan to brown. The comparative darker brown shade on the paper indicates higher level of concentration of lethal gas. It gives a fairly accurate reading when matched with calibrated color chart at lower concentrations. Thus in high concentrations this type has inherent disadvantage of infecting the crew before its presence can be selected. Therefore it is important to wear air mask.

Hand Operated Tube Detectors

This is one of the most accurate of manual detectors. It precisely measures concentrations levels from 0 ppm to 1000 ppm. It works by drawing H_2S through the sensor tubes filled with lead acetate coated Silica gel. This instrument gives the reading of concentration level of gas. It's also known as dragger tube.

Personal Electronic Monitors

The unit is an excellent piece of equipment supplying an accurate digital readout of H_2S concentration in parts per million (ppm). The units are usually hand held or belt mounted and measures the H_2S concentration at the sensor head continuously. Monitors give an audible alarm at a preset level of H_2S. The detector measures H_2S very accurately from 0 to 50 ppm. The warning—which sounds at level of 10–20 ppm.

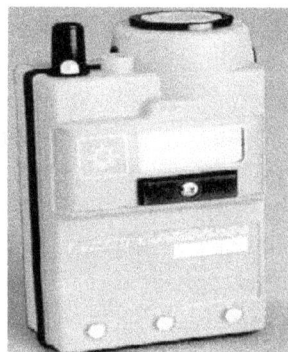

Portable gas detector

Fixed Detectors

This is most advanced piece of H_2S safety equipment in the offshore. This gives both visual and audible alarms. It uses several monitors or sensors heads placed at each strategic area. Which is susceptible to H_2S presence and accumulation. When the level of H_2S concentration reaches 10 ppm, the flashing lights appear and at level of 20 ppm or more, the lights and the siren are activated.

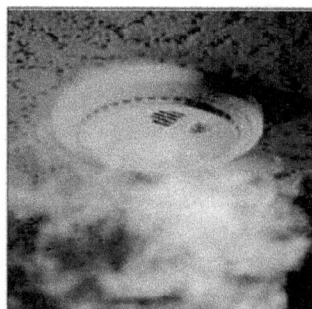

Fixed detector

5.3.3 Working Standards for H_2S Environment

An H_2S environment is defined by any location where the H_2S concentration could exceed 50 ppm. In those H_2S environment, following are the safety standards to be followed.

- All personnel on location shall be instructed on the characteristics and hazards of H_2S and the precautions necessary to assure safety.
- Every crewmember shall have self-contained breathing mask at all time. All other non-essential persons should be evacuated to the safe breathing area during operation.
- A primary and alternate breathing area is to be established so that at least one area will be free of H_2S depending upon on wind direction.
- Wind Socks shall be located at convenient location so that personnel are able to observe them.
- H_2S detection system shall be installed at various points on the location with one detector near work area. The detection equipment shall be equipped with both audio and visual signals.
- All casings, tubing valves, fittings, flow lines and surface handling facilities shall be made of material defined by API, ANSI or NACE standard of H_2S service.
- An H_2S contingency plan in the event of any emergency shall be known to all crewmembers.
- While working in probable H_2S contaminated area, the H_2S level should be monitored continuously.
- A well-defined briefing Area shall be designated on the platform for follow-ups in case of emergency and for regular safety drills.

5.3.4 Protection from H_2S

If working in H_2S environment, there are three categories of breathing equipment that are used to protect individuals from the dangers of H_2S exposure.

Escape Unit

Escape units are located near workstations. They have a small, self contained air supply and are designed to give enough air to reach safe location in event of an emergency.

Work Unit

Work units allow working for an extended period in an H_2S environment. They have an airline from a supplied breathable air source.

Rescue Unit

Rescue units provide a self-contained 30 minutes supply of air usually carried on back. They weigh about 35 pounds. Time can vary from person to person

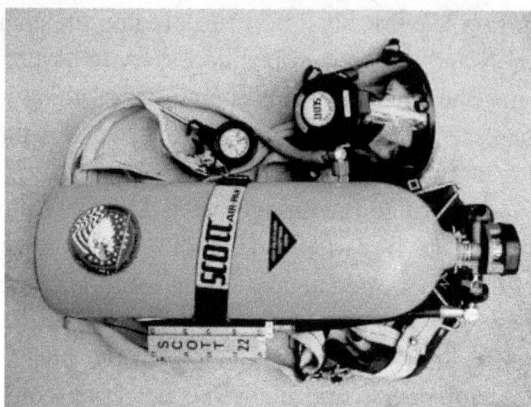

SCBA set

and also depending upon the activity, so never count the 30 min time. Audible alarms warn when air supply is low. After alarm sounds, one has 5 to 7 minutes of air left. Rescue unit may also be used as work units. Facial hair and absence of dentures could cause an improper seal, and also contact lenses should not be worn when working in H_2S environment.

5.3.5 Emergency Rescue under H_2S Exposure

Every person on board should be well educated and well briefed about the "emergency rescue" procedures o be adopted, if he finds that someone has been severely exposed to H_2S. These procedures are as follows:

- Put on proper rescue respiratory system.
- Move the victim to fresh air at once i.e. UPWIND or CROSSWIND.
- If victim is unconscious and breathing has stopped, apply mouth-to-mouth respiration immediately and continue until a resuscitator is brought in or normal breathing is restored.
- After reviving a victim, never leave him alone.
- To operate the oxygen resuscitator, first place blanket under the victim shoulders to open the airway. Open the oxygen air supply by turning valve on the top of supply bottle. Place the mask over the victim's nose and mouth and press the button to supply oxygen to his lungs. When victim's lungs expand, release the button so victim can exhale. Repeat this procedure at the rate of about 12 times per minute.

5.3.6 Location Safety in H_2S environment

Warning Signs regarding H_2S gas and its effects should be strategically located around the FPS. And each person on board must be aware of these signs and signals and the consequent dos or don'ts. Locational safety in H_2S environment must have the following:

- *Buddy System*—When H_2S reaches a high-risk concentration; workers should team together and work in pairs. The system is effective only if the workers stay together and are watching for early signs of H_2S poisoning.
- *Lanyards and Safety Belts*—If the distance between buddies must be extended more than arms length, a lifeline should be secured between them. The lifeline should be at least 400 lb test, soft fire resistant rope.
- *Cascade System*—It is a supplied breathing air system usually consisting of 360 cu ft compressed air bottles interconnected to provide breathing air to workers. The system is setup with a regulator to reduce the air pressure going to the work area. From a cascade system low-pressure hoses connect to manifolds into which each worker can connect the hose line for his work-escape unit.
- *Briefing Areas*—Each working station usually provides at least 2 briefing areas and are located on opposites sides of the location in order for one area to be upwind at all times. The upwind Briefing area is the protection center in the event of an H_2S emergency. These are the areas where rescue and work units in addition to other safety equipments are usually maintained.
- *Windsocks*—Wind will disperse H_2S very rapidly. Windsocks should be installed around the location for determining prevailing wind direction. All personnel on location should develop wind direction consciousness.
- *Bug Blowers*—Large blowers or fans may be used to disperse H_2S. In calm and extremely light winds, bug blowers are effective in reducing H_2S concentrations in the work area. Bug blower should be non-spark, explosion proof type.

5.4 EMERGENCY PLANS

In spite of all prevention measures specified and put in place on any floating production system, there remains an element of risk. Emergency procedures are therefore designed to indicate the duties, equipment and instructions so that necessary action can be taken in an uncontrolled situation. Personnel unfamiliar with emergency procedures can make the situation worse, where as those who are well trained can take the right steps to prevent it from getting worse. The procedures must be checked through training and exercises. If these exercises show that the personnel on-board FPS does not have the necessary capabilities and practices in the use of the equipment and procedures, more instructions and training will be necessary. The exercises also reveal, if the procedure is suited to the situation at hand. If this is not the case, the procedure should be revised. If the total approach to any given situation is correct and the guidelines are checked during exercises, that very knowledge will generate much greater confidence in personnel on-board to tackle any emergency situation.

Normally on any FPS, three types of emergency plans exist:

- Emergency Response Plans (ERP)
- Disaster Management Plan (DMP)
- Oil Spill Contingency Plan

5.4.1 Emergency Response Plans (ERP)

Aim of ERP is to provide Guidance to the personnel for action to be taken under various emergency conditions that arise on FPS or in a sea condition where FPS is deployed. Hence ERP is specific to a given floating production system. This document states responsibilities of individuals, departments, organizational resources available for use, sources outside the organization, general methods or procedures to follow, authority to make decisions, requirement for implementing procedures within departments, training in and practice of emergency procedures, communication and reporting required.

The re-course to ERP is taken under following situation:

- Hydrocarbon events-process leaks and fires, riser pipeline leaks and fires, blowouts
- Non-hydrocarbon events-ship collision, structural failure, helicopter accident, occupational accident
- Non-process fires—non-process hydrocarbon fires, electrical and living quarter fires
- Lightening around the platform.
- Cyclone
- Acts of sabotage

5.4.2 Disaster Management Plan (DMP)

Aim of DMP is to visualise possible emergency scenario that are likely to occur and to evolve a pre planned methodology carrying out various emergency combating plans so as to lay down clear cut procedure to ensure a effective rescue and rehabilitation operations. The DMP is concerned mainly with" salvage and rescue operation" and "containment and extermination of hazards". Other aims are:

- To train operating personnel by means of exercises and drills so as to make them well acquainted with the response actions such that these can be performed with greatest efficiency in minimum possible time.
- To minimize damage to environment during emergency.
- This plan also aims at immediate response to emergency events to prevent its escalation to a disaster.

The re-course to ERP is taken under following situation:

- Uncontrolled process leak
- Process fire explosion
- Riser failure
- Helicopter accident
- Blow out
- Collision
- Structural failure
- Natural calamity
- Man overboard

5.4.3 Oil Spill Contingency Plan

The Purpose of this plan is to outline procedures should oil spill occur in and around the area where FPS is deployed. This aims at minimizing loss of life, minimizing the damage to property and environment and also to ensure recovery of oil from oil spill around.

The scope of oil spill contingency plan includes all floating and fixed platforms, vessels and pipelines belonging to or hired by Operator for the purpose of exploitation of hydrocarbon resources within EEZ of a given nation.

Floating Production System: Process Facilities and Utilities Design and Technical Considerations

CHAPTER

6

Any floating production system deployed in the field normally comprises of the following three major components: A Hull, Various Topside Facilities and A mooring system. Choice of either of these components depends upon a host of factors like sea and environmental conditions, oceanographic factors, field parameters, process and equipment requirements, field deliverables … etc. We have understood some basics of mooring systems in the previous chapter. In the next three chapters, viz, chapters 4, 5 and 6, we shall be discussing in quite comprehensive way the various process facilities and systems, utilities, process support systems and safety and fire fighting system, aboard any floating production system as modularized topside facilities. Essentially, the concepts remains the same for both a floating production systems and for a fixed platform system with respect to these systems, sub-systems facilities, equipment, flow and logical sequences; but some variation do exist in specifications, arrangements and sizing commensurate with the characteristic of any floating production system.

Here in this chapter, some general design and technical considerations will be discussed which the pre-requisites for the topside facilities on any are floating production system so as to perform or deliver its optimal best on a floating production system.

Before I deliberate upon these considerations, I would like students to go through the Standards and Codes to be followed w.r.t the design of the facilities as tabulated below. I would advice them to develop as much understanding of these standards and codes by going through their individual sites/books/manuals as they can because ultimately only the depth of these understanding will help them in going for the design and engineering of systems/process/equipment on floating production system in better and realistic way.

STANDARDS AND CODES

Table 1: Codes and Standards

- American Bureau of Shipping (ABS)
- American Gas Association (AGA)
- American Institute of Steel Construction (AISC)
- American National Standards Institute (ANSI)
- American Petroleum Institute (API)
- American Society of Mechanical Engineers (ASME)
- American Society of Testing of Materials (ASTM)
- American Welding Society (AWS)
- Factory Mutual Research (FM)
- Institute of Electrical and Electronics Engineers (IEEE)
- Instrument Society of American (ISA)
- National Association of Corrosion Engineers (NACE)
- National Electrical Manfacturers' Association
- Occupational Safety and Health Administration (OSHA)
- Rules and Regulations of the Certifying Authority (the certifying authority has to be decided upon at a later date)
- Underwriters Laboratories (UL)
- United States Coast Guard (CG)

Table 2: Codes and Practices to be complied

ABS	Rules for Building and Classing Steel Vessels
ASME Section V	Non-destructive Examination
ASME Section VIII	ASME Boiler and Pressure Vessel Code Pressure Vessels
AGA	American Gas Association
	Gas Measurement Committee, Report No. 3
ASME Section IX	ASME Boiler and Pressure Vessel Code Welding Qualifications
ANSI B2.1	Pipe Threads Specifications, Dimensions and Gauging
ANSI B16.5	Steel Pipe Flanges and Flanged Fittings
ANSI B16.10	Face-to-Face and End-to-End Dimensions of Ferrous Valves
ANSI B31.3	Petroleum Refinery Piping
ANSI B31.4	Liquid Petroleum Transportation Piping System
ANSI B73.1	Specifications for Horizontal, End Suction Centrifugal Pumps for Chemical Process
API B73.2	Specifications for Wellhead Surface Safety Valve for Offshore Structures

API RP 2G	Recommended Practice for Production Facilities on Offshore Structures
API RP 2L	Recommended Practice for Planning, Designing and Constructing Helicopters for Fixed Offshore Platforms
API RP 14C	Recommended Practice for Analysis, Design, Installation and Testing of Basic Surface Systems on Offshore Production Platforms
API RP 14E	Recommended Practice for Design and Installation of Offshore Production Platform Systems
API RP 14F	Recommended Practice for Design and Installation of Electrical System for Offshore Production Platforms
API RP 14G	Recommended Practice for Fire Prevention and Control on Open Type Offshore Production Platforms
API RP 500B	Recommended Practice for Classification of Areas for Electrical Installations at Drilling Rigs and Production Facilities on Land and on Marine Fixed and Mobile Platforms
API RP 520	Recommended Practice for the Design and Installation of Pressure Relieving Systems in Refineries. Part I Design, and Part II Installation
API RP 521	Guide for Pressure Relief and Depressuring Systems
API RP 550	Manual on Installation of Refinery Instruments and Control Systems Part I
API Spec 2000	Specification for Venting Atmospheric and Low Pressure Storage Tanks
API Std 2502	Lease Automatic Custody Transfer
API Publ 2534	Manual of Petroleum Measurement Standards
AISC Publ M010	Manual of Steel Construction
AWS D1.1	Structural Welding Code – Steel
CG 259	Electrical Engineering Regulations
CG 320	Rules and Regulations for Artificial Islands and Fixed Structures
FAA 150/5390 IB	Advisory Circular – Heliport Design Guide
IEEE Std 45	Recommended Practice for Electrical Installations on Shipboard
ISA S5.1	Standard for Instrumentation Symbols and Identification
ISA RP 7.1	Recommended Practice for Pneumatic Control Circuit Pressure Test
ISA RP 12.1	Recommended Practice for Electrical Instruments in Hazardous Atmospheres
ISA RP 20.1	Recommended Practice for Specification Forms for Process Measurement and Control Instruments, Primary Elements and Control Valves

ISA RP 20.2	Recommended Practice for Specification Forms for Resistance Bulbs, Gauge Glasses and Cocks, Pressure and Temperature Switches, Level Switches, Steam Traps, and Drainers and Miscellaneous Instruments.
NACE Std MR-01-75	Material Requirement – Sulfide Stress Cracking Resistant Metallic Material for Oil Field Equipment
NFPA 10	Portable Fire Extinguishers 1978
NFPA 11	Foam Extinguishing Systems 1978
NFPA 12A	Halon 1301 Fire Extinguishing Systems – 1980
NFPA 16	Deluge Foam – Water Sprinkler and Spray System – 1980
NFPA 70	National Electrical Code – 1981
NFPA 72A	Local Protective Signaling Systems – 1979
NFPA 72E	Automatic Fire Detectors – 1978
OSHA Std 19CFR 1910	Safety and Health Standards General Industry
ILC	International Labour Conference Conversations 92 and 133
IMCO	• Code for Existing Ships Carrying Liquefied Gases • Crude Oil Washing Systems • Inert Gas Systems for Oil Tankers • Tanker Safety and Pollution Prevention 1978
IEC (92)	Regulations – Electrical Installations on Ships

Topside Facilities Aboard a FPS: Some Basic Consideration

Let's familiarize ourselves with some very basic considerations w.r.t Floating Production Systems (FPS). The process facilities and utilities proposed for any FPS must meet the following design requirements:

6.1 Modularization of the units on separate skids

6.2 The process equipment must operate satisfactorily on a constantly moving vessel

6.3 Space and Layout considerations.

6.4 Flow Assurance: Provisions for production of high pour point crude and storage of stabilised crude;

6.5 Corrosion considerations, as the crude contains a high content of both H_2S and CO_2

The three very important considerations are:

• Buoyancy must equal weight plus vertical loads from the moorings and risers.

• Space available must equal or exceed the space required for the functions to be performed

• Motions, station-keeping and stability must meet minimum criteria.

6.1 MODULARIZATION OF THE UNITS ON SEPARATE SKIDS

The particular interest from the point of view of processing and utilities facilities and considerations is the "Modularization" of topside facilities. That is to say, all topside facilities (process, utilities, accommodation, safety, firefighting etc) are fabricated in form of a "module" at a fabrication yard and then these modules are transported to the construction yard (where the hull of floating production system is being fabricated or converted) and then these modules are lifted and placed on the top of the hull deck. Thereafter all these modules are integrated as per the process, flow and utility sequences diagrams making these topside facilities as a comprehensive whole logically and otherwise. Normally 8 to 14 modules for top-side facilities are there on any floating production systems. The number and size of these modules are commensurate with field, process and equipment requirement and field deliverables. Most of the times, these modules are single deck, but now a days two-deck or three-deck modules are also quite common. Each major system or facility is treated as a stand-alone system or facilities and accordingly gets modularized. Each module is fabricated, tested, and delivered independently to the fabrication shipyard where these are assembled into a complete package ready for integration, commissioning, and start-up. Structurally, these modules normally have transverse framing so as to help in easy integration on the deck and also with each others. Though each of the modules deploys similar structural concepts, they do vary individually structurally depending upon a host of factors like given deck space, payloads …etc. Further, one important thing to understand here is that as lifting and placing the modules on a deck is a very serious and concerted exercise, normally the modules weight is kept within a critical value, i.e., a value that depends upon the available lifting capacity of the chosen fabrication yard. Or else, we have to break the individual heavy modules into a number of small modules and here we have to do a lot of precision work while assembling those smaller modules into a normal whole.

Modularization of the units on separate skids is an important consideration here as it reduces construction time. Each module gets equipped with all piping, wiring, cabling, instrumentation, etc., to maximize testing and commissioning of the facilities at the fabrication site before installing them on the vessels. It also enables modification work on the vessel to be executed simultaneously with the process and utility modules. Actual arrangement of the modules will greatly depend on the selected vessel and will be a compromise of several factors like Operating convenience, personnel and equipment safety; Allowable load of existing vessel structure; Effect of higher deck load and its distribution on vessel stability and so on.

On any floating production, we can have following modules for top-side facilities: Production modules (comprising of riser and flow line inlets and pipeline networks), process module (comprising of separation systems along with the individual oil systems, gas systems, gas lift systems, produced water systems, injection water systems and so on; might be there in form of individual modules); Utilities modules (like Oil dispatch and Gas compression modules, power generation modules); accommodation modules (from as small as for two people to as big as for 250–300 people); safety systems and firefighting modules. We shall be discussing all these modules in next few chapters.

6.2 THE PROCESS EQUIPMENT MUST OPERATE SATISFACTORILY ON A CONSTANTLY MOVING VESSEL

The very important consideration is that we need to ensure that the production equipment will continue to operate satisfactorily within the envelope of expected vessel motions. For this, we have to understand two things: what are the effects of vessel motion on the processing facilities and what are the means or the considerations to offset those effects of vessel motion on processing facilities.

6.2.1 Effect of Motion on the Process Facilities

Oil field process equipment, like separators, heat-exchangers and scrubbers, rely on gravity separation. This can be the primary separation of oil, water and gas or further refinement of these products in treaters. Gravity separation, by its nature, requires quiescence. Any movement of the equipment can upset the quiescence by producing turbulence in the liquid and waves at the interfaces. These waves also cause turbulences or eddy currents at baffles, and other obstacles. Of the six degrees of vessel motion, roll and pitch have the most significant effect on the design of process equipment. The experience suggests that processing equipment is to be designed to operate satisfactorily in vessel motions of 2-degree pitch and 12-degree roll, maximum, single amplitude.

Further, In general, vessel motion could cause a number of process control problems, like the few ones mentioned blow:

- Variations in level control displacer apparent weight caused by the effect of heave accelerations.
- Variations in liquid level resulting in liquid dump valve "hunting".
- Variations in liquid level causing nuisance high-and low-level alarms and shutdowns.
- Hang-up of floats or displacers with roll or pitch motions causing false control readings.

6.2.2 Considerations to Offset the Effect of Vessel Motions

Understanding the effects of vessel motion on process and controls, it is quite pertinent that while designing equipments, enough considerations must be given to avoid the tangible and intangible effects of the motion of vessel. Some pointers in this direction for considerations are discussed in following paragraphs:

- Separators should be equipped with anti-surge baffles to reduce the effects of splashing and sloshing. One should consider the use of double-barreled separators so as to reduce the effects of FPS motion. The use of curved weirs with a profile ensures that their flow characteristics are unchanged for all expected angles of roll. Separators should be located with their axes parallel to the longitudinal axis of the vessel. They should also be located near the vessel's center of gravity, where the motions are least severe.

- Level transmitters shall be placed in the centerline of the separator entering from the top of the central stilling tubes, from which all connections should be taken, to reduce effects of motion if float or displacer-type transmitters are used. Orifice plates can be used in the level connections to damp wave motion; and triplicate level switches, on a two-out-of-three voting system, should be provided for low and high level shutdown, to avoid spurious trips.

- The common float-type level controllers will interpret the movement of an interface as a change in the liquid volume to be corrected by adjusting the rate of liquid leaving the separator. Furthermore, the vessel motion will induce forces directly on the displacer, causing it to bob bump inside its stilling well or containment tube. These movements will likely be interpreted as if caused by an actual liquid level change. The result could be a constant opening and closing of level control valves and, hence, poor flow control into the separator. In certain critical control loops, poor process performance and instability could result.

- To minimize these problems, the use of conventional float-type level controllers should be avoided unless they are specially designed for marine separators. Instead, differential pressure (d/p) sensors should be used. The d/p-type sensor uses a thin diaphragm to sense the average height of liquid rather than the actual interface position. As a result, if the sensor is located at the appropriate point in the separator, the motion of the interface will be ignored by the d/p-type sensor. For sensing oil/water interfaces, displacer-type instruments with marine guiding give good results. We can also use as an alternative method the isotope-type level transmitter.

6.3 SPACE AND LAYOUT CONSIDERATIONS FOR EQUIPMENT

The equipment or the top side facilities on the deck of any FPS follows a very specific space and lay-out consideration. There are certain factors that must be put into consideration while considering the space/area allocation for equipment. These factors are:

* Operational risk
* Safety, classification of hazardous areas, and equipment grouping in accordance with API RP2G, ABS requirements
* Operations and maintenance requirements
* Weight distribution
* Navigational requirements for helicopters
* Process flow and work-over considerations
* Future facilities

Top side facilities are arranged from the point of view of operational risks and as per this consideration; facilities are arranged in decreasing level of operational risks from bow to stern side of the vessel. On the Floating Production System, the very first important consideration is to locate the process equipment as near to the center of gravity of the vessel as possible, where the vessel motions are the least severe. Hazardous facilities are located away from the accommodation module. The longitudinal axes of separators are to be oriented in parallel with the longitudinal axis of the vessel to reduce the effect of roll motions.

For safety reasons and to meet code requirements, the process facilities should be located as far as possible from any ignition source. In a well-designed layout of the topside facilities, hazardous facilities such as boiler and generators are located in a common utility area and at a safe distance from the crude processing facilities. In addition to the area and space requirements of major equipment, the distribution of the weight of these facilities on the deck is also a prime consideration in determining the final layout of the topside facilities.

The location of the living quarters and helideck is dictated by codes and regulations (e.g. API RP2L). The API requires that the helideck has a completely unobstructed approach departure angle of not less than 180 degrees when viewed in the plan. The remaining sector must contain no obstructions within a specified radius exceeding a specified height above the helideck. This radius and height is determined by the type of helicopter used. In relation to the helideck, the facilities on the deck should be arranged as low as possible. As the storage tanks are located below deck, the crude loading pumps, stripping pumps, ballast pumps, and seawater pumps have to be installed at the same level as the bottom of the storage tanks due to Net Positive Suction Head (NPSH) requirements. In the semi-submersible FPS case, the work-over (or drilling) mono-pool dictates the location of the work-over (or drilling) rig and the pipe lay down area. The work-over rig will typically be located at the center of the semi-submersible. Finally, some space should be allowed on the FPSO for future facilities.

6.3.1 Layout of Equipment on FPS

Just to develop an understanding, let's understand how equipment is being laid out on FPSO Barge and FPSO Tanker.

FPSO Barge

At FPSO Barge, a mono-pool is located at mid-ship to facilitate wireline operation. The helideck is located at the bow and provides a completely clean landing area. The 24-man living quarters and the control room, made up of portable cabins, are located below the helideck. The process equipment and meter prover are mounted on a platform located near the mono-pool. This location is close to the center of gravity of the barge, where the vessel motions are the least severe. The platform is raised 3 meters above the top deck for ease of operation and maintenance, and to avoid waves splashing on equipment and also as required by codes and regulations. The generators, boiler, and other utility features are also mounted on a platform at the stern. The inlet manifold is located at starboard. A 20-ton hydraulic crane is provided at port side for handling of supplies and maintenance. The pump room is located below deck, aft of all cargo oil tanks. It accommodates the offloading, stripping, seawater, ballast, and one of the fire pumps. The other fire pump is located on the deck to give a wide separation of these emergency facilities.

FPSO Tanker

At FPSO Tanker, the process equipment is mounted on a platform located at mid-ship. This location is close to the center of gravity of the tanker, where the tanker pitch and roll motions are least severed. The platform is raised 3 meters above the deck for ease of operation and maintenance. The meter prover is also mounted on this platform. A helideck is located just forward of the existing deck house. The ideal location of the helideck will be at the stern. However, the rigid arm extension attaching the tanker to the turret mooring column occupies most of the space at the stern. It is not practical to raise the helideck above the rigid arm. A ground flare is located at the forward section for disposal of excess gas. Two pedestal-mounted cranes are located between the process platform and the ground flare for handling of materials from the supply boats and also for lifting the equipment during maintenance. A gantry is provided on the forecastle deck for handling of loading hose since tandem-to-tandem offloading is contemplated. The machinery room is located below deck, under the deck house. Existing boiler, generators, offloading pumps, seawater, ballast and fire pumps, etc., are located in the machinery room.

6.4 FLOW ASSURANCE: PROVISIONS FOR PRODUCTION OF HIGH POUR POINT CRUDE AND STORAGE OF STABILISED CRUDE

High pour point crude presents a number of production and operational problems. The production of high pour point crude can result in wax deposition inside the well bores, production facilities, pipelines, and storage tanks. These wax-deposit problems can vary from very minor to extremely severe, depending on the wax content of the crude oil, the cloud and pour points of the crude, and the operating temperature. When waxy crude is cooled, the wax gradually crystallizes in the form of thin plates or needles. When enough wax has crystallized, it can form a three-dimensional network throughout the crude and cause solidification.

Much of the early research on the flow properties of high pour point crude was done to perfect the methods to transport this crude below their pour points. In order to produce waxy crude from the field, its fluidity in the production system must be maintained. The fluidity of such crude is usually maintained in the production system by insulation and addition of heat.

The pour point is often used an indication of a crude oil's flow properties. Above the pour point, the crude typically behaves as a Newtonian fluid. However, as it cools, it becomes non-Newtonian; therefore, its viscosity varies as a function of shear rate (flow rate in a pipe). Recently, it has been recognized that other flow properties of crude, such as their gel strength and viscosity, should also be considered together with their pour points.

High pour point crude have been successfully produced from oil fields in China (Daquing, Shengli, and Bohai Gulf), India (Bombay High), Indonesia (Udang), the North Sea (Beatrice), the United States (South Louisiana and Utah), and Russia. A wide range of solutions has been developed to solve the operating problems caused by wax deposition. Many techniques, chemicals, and devices have been used with fair amount of success.

For producing high pour point crude, following design philosophies be considered/ adopted:

- Maintain the fluidity of crude by keeping the oil hot.
- Provide a backup system to restart the flow of crude should the primary preventive method fail.
- All surface facilities handling high pour point crude should be insulated to minimize heat losses.
- All process piping should be heat-traced. This will facilitate the restart of congealed crude in the piping after a prolonged shutdown.

One of the major operational problems in producing high pour point crude occurs during the start-up of new facilities. The production facilities have to be heated to process temperatures (about 42°C) prior to flowing through the production system. One solution is to flow hot fuel through the process train before producing the crude. This is done in the Udang field. From an operational point of view, it is always good practice to wash out residual high pour point crude with diesel oil from the processing facilities before shutting them down. This practice should also be followed prior to shutdowns for routine maintenance or repair. This will facilitate quick start-up. When there is insufficient time to displace all the high pour point crude from the production system, quick start-up of the facilities may be facilitated by injection of a compatible pour point depressant prior to shutdown.

Another concern in the design of facilities for production of high pour point crude lies in pump selection. Pumps are required for crude transfer from the storage to the tanker. It is noted that as flow decreases from some given optimum point, the pressure requirement increases. This is because of non-Newtonian behavior; the longer the oil is in the line, the more it cools and the more its viscosity increases. When the flow starts to increase past the given point, Newtonian characteristics prevail and pumping pressure increases due to higher flow. In choosing a pump for high pour point crude, care must be taken to ensure the operating point Pmax (maximum pressure), Fmax (maximum flow) is met. If restarting pressure is greater than maximum pressure, then special start-up pumps, such as positive displacement pumps, must be considered.

6.4.1 Improving Fluidity through Heating Options

Heat is required to maintain the fluidity of the waxy crude, especially in the storage tanks and crude processing stream. Normally all FPS has duel-fuel boilers onboard that generates steam which acts as a heat source. Additional heating requirements are supplemented by recovery of waste heat from exhaust of gas turbine, exhausts of compressors and generators, exhaust of diesel engine, from boilers and from the ground flare. Optimizing operations and cost savings can be achieved when waste heat can be used. Apart from heating waxy crude, waste heat can also be used for producing steam as well as heating the living quarters.

Boilers can be designed with such high efficiency that recovery of the remaining waste heat is not sensible. It is, therefore, recommended to install high-efficiency boilers equipped with combustion air preheaters in the exhaust stack. Inert gas generation can be combined with waste heat recovery; however, economics need to be worked out carefully. Gas turbines are a good source of waste heat and may also be a continuous supply of heat, depending on its application. Diesel engines cooling systems are a source of low-level heat; and if there is an application for it, it is a valuable source (for instance, space heating). The exhaust gas of diesel engines may also be a good source of recovery heat.

6.4.2 Drag Reduction in Fluid Flow and Flow Assurance

Drag Reducer is a flow improver which increases the flow capacity. Certain additives known as Ddrag Reducing Additives (DRA) are added to improve flow in pipelines by reducing turbulence. They can dramatically increase flow using minimal additional of energy, or they can sustain a given flow rate using less energy.

In most petroleum pipelines, flow is turbulent. Non-linear currents and friction cause much of the energy applied to move the fluids to be wasted. Drag Reducers are long-chain hydrocarbon polymers that reduce friction near the pipeline wall and within the turbulent core, dampening rotational flow and thereby decreasing energy loss.

Turbulent drag reduction in fluid flow by additives has been researched since quite long, probably since 1949. Various technical applications have been envisaged both for polymeric and surfactant drag reduction. Some successful applications include increased flow of oil and other fluids through pipelines, improved fire fighting and irrigation, improved flow in storm sewers, and flow augmentation by drag reducers in pipeline transport of sediments and slurries. Other possible applications include district heating circuits, industrial and agricultural sprays, and slow-release fertilizers and pesticides. However, the mechanism of drag reduction still eludes exact explanation mainly because of the inability to characterize polymers, surfactants, and their interactions at molecular or micellar levels in turbulent flows.

6.5 CORROSION CONSIDERATIONS

If the crude contains a high concentration of H_2S (>230 ppm) and CO_2 (>3%), corrosion protection becomes very important.

For any oil industry man, it is important to have sufficient knowledge of corrosion mechanism and its remedial measures. A brief discussion has been made at Annexure H. However, it is advised that a good understanding be developed over this important concept through other books.

Sub-Sea Production Systems

7

Sub-Sea production systems refers to a comprehensive production system wherein equipment, systems and sub-systems for producing oil and gas and its processing (though not in use widely currently; still emerging) are situated in the vicinity of the seabed for the purpose of hydrocarbon production. Though a dream in 1960s, this sub-sea has now come in a long way and currently a number of fields at greater depths are simply and reliably producing through sub-sea production system. This is a more appropriate and a cost-effective technique that has taken over the use of fixed platforms at sea for the fields at ever-increasing depth. It eliminates the cost of lifting hydrocarbon fluid to the surface and hence eliminates the need for a fixed platform. Besides, this is a better and safe option for fields located in the extreme and harsh climatic sea conditions.

Sub-sea systems, in general, comprise of multi-component seafloor facilities that allow production in water depths that would normally preclude installing conventional fixed or bottom-founded platforms. Sub-sea systems are divided into two major components: the seafloor equipment and the surface facilities. The seafloor component includes one or more sub-sea wells, well heads, manifolds, control umbilical and flow lines. The surface component includes the control system and other production equipment connected by gathering lines from the sub-sea wells and installed on a host platform in shallower water set many miles from the actual wells.

Sub-Sea Production System Comprises of the following components:

7.1   Sub-Sea Templates
7.2   Sub-Sea Manifolds
7.3   Sub-Sea Completion
7.4   Sub-Sea Control Systems
7.5   Sub-Sea Flow Lines and Risers

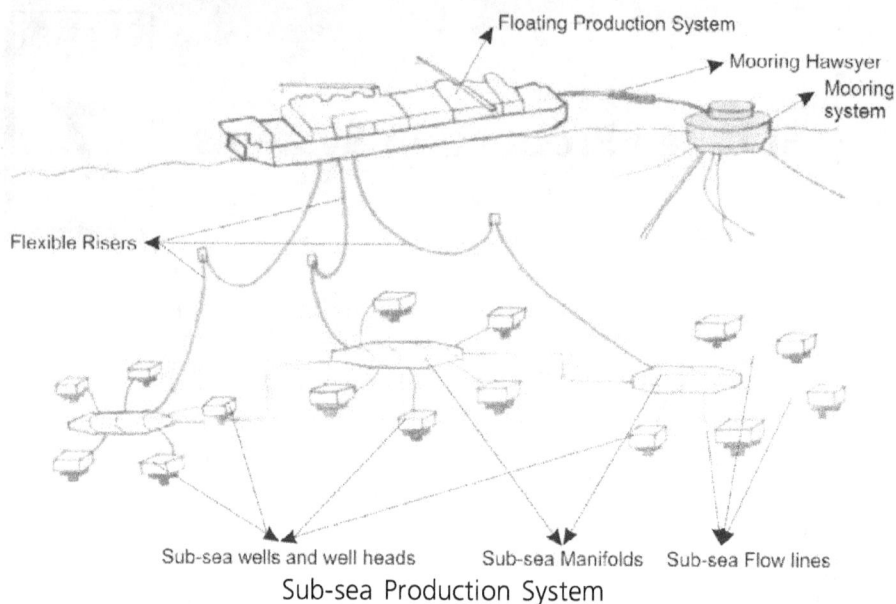

Sub-sea Production System

7.1 SUB-SEA TEMPLATES

Sub-sea template is a space frame enclosing all sub-sea production equipment inside an open framework of structural members. Sub-sea template structures support several sub-sea wells from a common base. The structure positions the well slot and manifold to optimize the structural size and configuration. It also provides a guidance system (guideline) for drilling and completion operations. The structure is designed for installation by a crane barge and incorporates a mud-mat support and hydraulic leveling system for leveling the structure prior to pilling to the seafloor. The template structure has a specified number of well slots arranged symmetrically on either side of a central manifold. A pipeline and umbilical connection area is provided on one end of the template structure to facilitate diver-assist sub-sea line connections. The template structure is normally specified by its length, width, height, weight and well spacing.

The principal advantage of use of templates in sub-sea production results from some or all the wells being in one location, thus simplifying many a functions, in particular that of flow line needs. They also permit direct vertical access into all parts of the template from a single overhead location. Further, the template may act as a base for the production/export riser and take both the vertical and horizontal forces which arise from the movement of the FPF and any flow line pull-in operations or snag loads on the attached flow lines. Template also supports pipe works, valves and hydraulic operators which are retrievable for servicing and replacement.

A template may simply be a straightforward drilling guide framework or it may be a more complex pre-engineered assembly of wellhead, flow line, manifold and control hardware. The important consideration here is that in all the cases, the template must have a simple but reliable leveling system in order to achieve the required tolerance for drilling operations. The template may be fixed on location by means of piles for leveling prior to drilling.

A production template is made up of a drilling template, the production equipment and control equipment. Drilling templates can be modular or unitized. The production and control equipment consists of the manifold, trees, and control system. The manifold and the control system can be partly or fully modular to be integral with the drilling template structures. With a cluster of wells on an integral template structure, the drilling rig maintains a single station to drill deviated wells. Original exploratory wells or new wells can be routed to the template as satellites. An FPF (fixed platform) sited vertically over a template is generally designed to carry work-over capability. However, such activity may require a production.

The most common types of sub-sea production templates are as follows, primarily depending upon the kind of flexibility they provide with respect to subsequent well additions should the reservoir so dictate in her later phases:

7.1.1 Unitized Template with integral manifold
7.1.2 Unitized Template with modular manifold
7.1.3 Modular Template with modular manifold

7.1.1 Unitized Template with Integral Manifold

The unitized template with integral manifold is the least flexible option. Here the number of wells is fixed. Here templates and manifold are fabricated together. Further, as the manifold must be completed prior to installation, the amount of time available for pre-drilling is minimal. Unitized templates with integral manifolds have some disadvantage when it comes to replacement or repair of components on the built-in manifold. Therefore, special "insert" valves have been developed for replacement through remotely-operated maintenance vehicles.

7.1.2 Unitized Template with Modular Manifold

In this type of templates, manifold modules can be fabricated independent from the template itself. This ensures much more flexibility, both in terms of the overall project schedule as well as the choice of fabricators. The use of a modular system on a template allows early installation of the template and enables drilling to commence during the construction of the manifold prior to the arrival of the FPF (Fixed Platform). This pre-drilling enables a rapid build-up of production once the FPF is on station. The manifold may be installed just prior to the FPF coming on stream.

A modular type of manifold provides the capacity to carry out maintenance and any changes to the manifold on the surface that may be necessary due to the changes in well function during the life of the field. In addition, certain maintenance-prone components, such as frequently used valves and chokes, may be designed for individual retrieval, independent of the manifold module. However, this has some risks relating to the mechanical fit-up and related tolerances.

7.1.3 Modular Template with Modular Manifold

A fully modular system, including a modular-type template, is normally considered when a limited number of wells are to be drilled. This type of system is meant to give complete flexibility; flexibility in the sense that extra wells may be added relatively easily should the reservoir so dictate. Barring this one advantage, this option has a number of disadvantages. The trees chosen with this option have shared guide posts that require sequenced removal of complex guide funnel systems for maintenance which is a complex and costly exercise. Also, it is difficult to successfully make the multiple vertical stabs which are required to connect all the trees. Further, it is experienced that the capital cost savings over a unitized template are not significant.

One important thing to understand here is that the different kind of sub-sea templates as discussed above follows a definitative sequence with respect to installation and drilling. The sequence must be crafted carefully to avoid delays and complexities and to ensure savings on cost and time.

7.2 SUB-SEA MANIFOLDS

The sub-sea manifold provides the flow path between the template wells and the sub-sea pipelines. It combines the production flow lines from all template wells into a production header. Sub-sea manifolds are used to reduce the complexity of risers, flow lines, and control system umbilical, if a large number of wells are involved. The manifold is integral with the template structure and is designed to be diver-maintainable.

In any manifold system, it is preferable to include service and test headers which will enable individual wells to be serviced or tested without interruption of the main production stream. A test header is provided at the manifold to enable individual well production to be diverted to the FPS for well testing. The test header is of the same size and is looped with the production header. The test header is also used for round-trip pigging with the pigs launched and received on the FPS, as well as providing a redundant production flow path to the FPS. An annulus header is provided on the manifold and is tied into all the template wells' annuli. The annulus system enables constant well annulus pressure monitoring and bleed-down. The annulus piping is normally crossed over to the production piping on the tree assemblies. This arrangement enables flushing and testing of the production piping and flow control choke.

To balance the flow from or into a manifold, special adjustable chokes are used. The chokes are special because it is different from the one currently used at surface operations. These chokes are either remotely adjustable through ROV (remote operated vehicles) or manual adjustable by divers and have a long, trouble-free service life, thereby reducing the costly sub-sea maintenance operations. Any sub-sea manifold requires a planned choke maintenance schedule in place because choke wear is quite prominent here requiring timely calibrations but unfortunately this choke wear are less visible or less indicative. Further, we need to have an improved and reliable instrumentation of the control system for monitoring the flows and pressures in a manifold.

All pipeline and control umbilical connections to the FPS are located at one end of the manifold. Placing the line connections at one end provides a means to optimize line routing to the FPS and provides sufficient working area to permit divers to make up the line connections.

If it is required to connect a number of different facilities (satellite and/or templates) to a floating facility, then the maximum amount of manifolding commensurate with flow line/riser etc. is normally placed below the FPF (fixed platform). This enables maintenance of those manifolding-related equipment like the chokes to be carried out without the mobilization of separate service vessels.

7.2.1 Sub-Sea Manifolds: Types and Common Attributes

There are basically three types of sub-sea manifolds depending upon how manifolding is done sub-sea:

- Satellite manifold centers
- Template manifolding
- Riser manifolding.

The different types of a manifolding system have some or all of the following attributes: The sub-sea manifold

- Commingles well production into common headers and controls individual well flow rate control (choking).
- Provides access to well annulus for annulus monitoring, down-hole injection, and well killing
- Provides maintenance capabilities for line flushing, inhibition injection, and paraffin cutting.
- Provides a header system for gas lifting of wells, including individual well flow rate control and metering
- Provides TFL (through tubing flow line) tool diversion to wells.
- Provides a system to detect and limit sub-sea pollution from manifold leakages

7.2.2 Sub-Sea Manifold: Advantages and Disadvantage

Sub-sea manifold has a number of advantages. These manifolds at sub-sea lead to lower riser hardware and hence lead to lower installation costs. Further the presence of manifold gives the less interference between lines of the risers. Besides this, future wells can be tied into the manifold with minimum effect on the riser or FPF.

However, the sub-sea manifolds present some areas of concerns that need to be addressed adequately. Sub-sea manifolds leads to expensive manifold and sub-sea controls besides giving longer lead time for manifold delivery. One single failure at sub-sea leads to have greater impact on operations. Sub-sea manifold with wet trees poses a lot of problems. The flow line connection sometimes leads to costly operational delay risks because wet system repairs usually take longer time and require larger equipment. Further, it requires the rig for running and manipulating remote connectors and this is a costly exercise. And besides this, detection and repair of minor leakage is difficult with wet manifolds. This is an important concern in sub-ice systems where leakages don't disperse.

7.3 SUB-SEA COMPLETION

A sub-sea completion system is basically an assembly of individual subsystems which can be utilized in many forms to suit the development requirements of a particular field. In general, sub-sea completions are of two types: either we can have Template manifold sub-sea X-mas tree or we can have a stand-alone sub-sea X-mas tree.

A sub-sea completion system normally comprises of the following, inclusive of their own running/testing/handling tools:

7.3.1 Wellhead Completion System
- Guide bases
- High-pressure wellhead housing
- Tubing hanger

7.3.2 Tree Assembly
- Wellhead connector
- Valve assembly
- Choke assembly
- Tree cap
- Tree flow line, flow line connector and control

7.3.1 Wellhead Completion System

A well-head completion system for a sub-sea completion comprises of a guiding base, high pressure wellhead housing and tubing hangers.

Guide Bases

The guide bases provide guidance for the drilling, casing strings, BOP (blow out preventor), and ultimately to the production tree assembly. That's why; it is named as guide base. Guidance onto the support base is by means of four surface-tensioned guide wires which are attached to guide posts on the support base itself. We have two types of guide bases. The most common is to use a standard exploration-style drilling support base (also called a permanent guide base) for single (satellite) wells which form part of a sub-sea production system but are located in remote areas of the field. The other one is the use of a template structures as a common support base for several wells, which allows for a more rapid drilling programme due to the virtual elimination of rig move times.

High-pressure Wellhead Housing

The wellhead housing system seals and supports the casing strings, large sub-sea BOP stack, the tubing hanger and production tree assembly. The wellhead hosing commonly used are either a two-stack system say, of 20–1/4 inch × 2000 psi and 13–5/8 inch × 10,000 psi wellheads, or single-stack 18–3/4 inch × 10,000 psi system. However, due to the size and weight of the two-stack system or of the 18–3/4 inch sub-sea BOP stacks and marine riser systems, the use of the more compact 16–3/4 inch × 10,00 psi system on the floating production vessel has emerged as a better option so as to minimize deck loads and reduce handling problems. The diameter and pressure figures given here is indicative only to develop some understanding.

Tubing Hanger

The tubing hanger system runs inside the sub-sea BOP stack and seals and lock down inside the high-pressure wellhead assembly. Commonly, a single tubing string is suspended from the tubing hanger in the sub-sea well, with provision in the hanger for annulus access and the hydraulic control of down-hole safety valves. Profiles in the tubing hanger bores allow wireline plugs to be installed for the safe removal of the BOP stack before installation of the production tree assembly.

Installation of tubing hanger is an important factor for proper place-up. Two installation methods are currently available to run and control the tubing hanger system. The "mechanical system" requires two different tools to install the hanger. The first tool is used to run, lock and seal the hanger in the wellhead whereas the second or tieback tool allows vertical access through the tubing hanger and into the production tubing string. As this "mechanical" system requires at least two trips from the rig to the wellhead therefore, this option is considered for use only on smaller sub-sea completion systems. The other system is the "hydraulic" system which requires only one trip to install the hanger and tubing string. The running tool is controlled by a six-function hydraulic umbilical. A further development on "hydraulic" system front is that a completion riser system has been developed for use with the hydraulic set tubing hanger which reduces the installation times to a minimum and also helps in to install and work-over the production tree assembly. Although the basic hydraulic equipment is more expensive than the mechanical system, the rig time saved by the elimination of additional trips gives the hydraulic system more economical method on the larger field developments.

7.3.2 Tree Assembly

The trees are identical assemblies, consisting of a master valve block, wellhead connector, valve assemblies, tree cap, tree flow line systems, flow line connectors and associated controls, flow control choke, and annulus and production piping. The primary purpose of the trees is to provide primary well control and production flow control. The tree assemblies are installed as a module after well completion via a chosen FPS. Divers are utilized to make the tree piping connections to the manifold. The tree assembly of a sub-sea completion system comprises of the following:

Wellhead Connector

The wellhead connector is designed to lock and seal the production tree assembly on the high-pressure sub-sea wellhead housing. Basically this wellhead connector's top face acts as a base for the production tree valve assembly. An integral hydraulic system allows the connector to function while a ring gasket, carried inside the connector, provides sealing inside the wellhead. A series of locking dogs is designed to engage an external profile on the wellhead housing and to set up the preload required to energize the ring gasket. The locking dogs are activated by means of a cam ring which is connected to the integral hydraulic system.

Valve Assembly

The valve assembly is mounted above the wellhead connector and is designed to control and direct the produced hydrocarbons into the flow line system. There are seal subs in the base of the valve block that extend into pockets in the tubing hanger so as to isolate the flow from the production tubing and well annulus. Further, the master block of the valve assembly has a series of fail-safe gate valves to control the fluid flow and to allow vertical access through the block during installation and work-over operations.

Tree Cap

In its basic form, the tree cap simply acts as a protective cover for the re-entry mandrel. It locks onto the outside profile of the mandrel mounted above the valve assembly. The tree cap is either mechanically operated or hydraulically operated. The tree cap can be used to provide a backup seal to the upper (swab) valves on the master valve block. Further, the cap may form a part of the production tree control system by housing the necessary control valves, filters, accumulators, etc. Using the tree cap to carry the control equipment eliminates the requirement for a separate control module and makes the system more compact, simpler and cheaper. However, it is desirable to have a separate control module, mainly because of the possibility of damage to the seal areas in the re-entry mandrel during the more frequent make-up and removal sequences. Further, some sub-sea tree caps do have a "trash cover" on it to protect delicate parts of the tree assembly against dropped objects.

Tree Flow Line System, Flow Line connector and Control

The tree flow line system comprises of two parts: the flow lines on the tree itself and the flow line connection between the tree and the production facility. Flow from the valve assembly to the flow line is through a side-mounted gate valve known as the wing valve. The connection between the flow line on the tree and the line to the production facility can take many forms, the simplest being providing of a conventional API flange through the assistance of a diver. However, more sophisticated systems are also there like a hydraulically-operated vertical stabbing connector or a fully diverless pull-in and remote make-up connection.

7.4 SUB-SEA CONTROL SYSTEMS

Sub-sea control system has two components: The one on the surface of the floating production systems like the hydraulic power unit (HPU) and Production Control Panels; the second component comprises of the equipment at sub-sea like Umbilical and Tree-mounted equipment. These together form a comprehensive sub-sea control systems integrated logically through a number of varied sub-sea-control logical options, the choice of which depends upon the sub-sea production system configuration, lay-outs and other considerations like type of floating production system chosen.

7.4.1 Hydraulic Power Unit (HPU)

The Hydraulic Power Uunit (HPU) is skid mounted and is normally designed for water-based hydraulic fluid. This HPU is installed on the main deck of the barge, in case of a barge-based system or it is installed on the multi-pass swivel of the SPM (single point mooring) of the tanker, in the tanker-based system. This HPU can, however, be placed on the tanker itself, at the expense of additional hydraulic lines passing through the yoke. But then, in the tanker-case, placing HPU on tanker is slightly less reliable than the option when HPU is placed on swivel. In both the cases, however, hydraulic fluid tank is on the main deck itself. The HPU is interconnected with the ESD of the floating production system so as to cause automatic shutdown of the HPU in case of an emergency if so arises and thereby shutdown wells sub-sea.

HPU generates hydraulic pressures to operate hydraulic actuators on the tree and for Surface-Controlled Sub-sea Safety Valves (SCSSV). The hydraulic pressure can be of different values depending upon the chosen configuration of tree assembly and SCSSV. Each hydraulic supply is powered by redundant air-driven hydraulic pumps having specified rating commensurate with HPU. Filters and accumulator banks are provided for each such supply. The normal practice is to have the accumulator banks of such sizes that operate all sub-sea valves twice.

The HPU has a reservoir to contain supply of water-based hydraulic fluid for a specified minimum number of days, say a minimum of 30 days' supply. Further, the reservoir is provided with an elastomer air bag in contact with the fluid surface to prevent contamination of the hydraulic fluid. The hydraulic fluid consists of 2% proprietary additive (a lubricant, a corrosion inhibitor, and a biocide) and the remainder is fresh de-ionized water and is preferably mixed onshore.

7.4.2 Production Control Panel

A self-standing production control panel for each well is placed preferably in the control room of the barge or tanker. It can, however, be placed outside the control room with remote control and monitoring features added to the panel so as to operate from control room.

These control panels are mounted in a unitized carbon steel skid. Each panel contains manual control valves to operate the X-Mas tree actuators. The control valves are arranged on a mimic panel which clearly indicates the flow of produced fluid in the Christmas tree. To indicate sub-sea actuator action, each line has re-settable, totalizing flow meters to measure total hydraulic fluid flow in either direction. Because the actuator volume of the SCSSV is very small, flow meters cannot be used to reliably indicate the action of the valve. The only other instrumentation on the panel is a pressure gauge to monitor annulus pressure.

The control panel has normally the following operating functions: Master valve (open/close); Wing Valve (open/close); Annulus Wing Valve (open/close); SCSSV (open/close); Chemical Injection Valve (open/close).

7.4.3 Sub-Sea Umbilical

Sub-sea control systems are exercised through umbilical. Umbilical is defined as a pipe comprising of a number of reinforced hoses connecting sub-sea wells to surface structures. It may contain electrical conduits, power supply cables, fiber-optics, hydraulic tubes, other pipes, and so on. It is used to operate and maintain sub-sea wells or other sub-sea equipment. Besides this, umbilical is also used for data and communication purposes (through fiber-optic elements) as well as for the purposes of controlling and injecting chemicals (like corrosion inhibitors, hydrate inhibitors) and for similar other purposes like supply of diesel fuels between two very nearby offshore platforms.

The hoses are arranged in a circular pattern, and the assembly is armored with a double helix of wire. Each layer of polyurethane isolates the wire from seawater and protects against abrasion. All hoses are rated for a specified working pressure, can have different internal diameters and the materials are compatible with the hydraulic fluid. The connections of the umbilical to the tree are all made up at once by the diver through actuation of lever arms in the connection plate. Connections between the flexible hose assembly are similarly made up.

In general, one hose each is provided for Production Master Valve (PMV) control, production wing valve (PMV) control, Annulus Wing Valve (AWV) control, annulus pressure monitoring, chemical injection valve (CIV) control, chemical injection fluid line and one hose is kept as spare. Further, two hose-lines are provided for SCSSV control because of it being the balance type.

7.4.4 Tree-Mounted Equipment

Tree-mounted equipment consists of the tree half of the umbilical connector, ½-inch OD stainless steel lines from the connector to each valve actuator (each line has an isolation valve) and isolation valves to connect the spare line to each other line.

The swab valve actuator is also connected to the tree half of the umbilical connector and is dead-ended at this point by the umbilical half of the connector. Further, there are hydraulic lines from the tree connector to the tree half of the hydraulic connector which is actuated, as required, by the work-over system for installation and removal of the tree. Whenever a work-over is required, the control umbilical is disconnected by divers and the work-over umbilical is connected in its place. Further, we have the manual valves that essentially serve two purposes. It permits easy replacement of a failed umbilical line by the spare line. Until that unlikely failure occurs, it can be used to flush the lines after initial installations and subsequently afterwards as the need arises.

7.4.5 Sub-Sea Control Systems Logic Options

The various components of sub-sea control systems are held together as a logical comprehensive whole through a number of varied sub-sea-control logical options, the choice of which depends upon the sub-sea production system configuration, lay-outs and other considerations like type of floating production system chosen.

The various sub-sea control system logical options are as follows:
- Direct Hydraulic
- Discrete Hydraulic
- Sequential Hydraulic
- Electro-Hydraulic
- Multiplexed Hydraulic
- Ultrasonic Control
- Hydro Kinematics Control
- Electrical Control

Direct Hydraulic

This is the simplest control system where each sub-sea valve actuator is provided with hydraulic fluid directly from the surface and all control valves are located at the surface control panel. As the direct hydraulic doesn't have much sub-sea active devices other than the actuators, it results in high reliability and significantly reduced maintenance costs. Operational flexibility is limited only by the capabilities of the surface control panel. However, as the distance from the production facility increases, increasingly larger diameter lines are required to maintain response times. And for these large distances, water-based hydraulic fluid is used for direct hydraulic control knowing that the flow rate is proportional to the viscosity of the hydraulic fluid. In this type of control, capital costs are normally higher since the control line bundle must contain one hydraulic line for each valve to be actuated. This type of control system is preferred in cases where distance to the surface control panel is less than or equal to 2 kilometers and response time on closing is less than one minute with 3/8 inch ID hoses. Otherwise, sometimes we need to incorporate boosters in the system.

Discrete Hydraulic

Discrete hydraulics utilizes individual control lines to operate pilot valves, which, in turn, admit hydraulic fluid from accumulators to the valve actuator. Fast response time during closing and opening is achieved with relatively small diameter control lines at distances of several thousand meters. The supply line need only be sufficiently large to recharge the accumulator at the desired valve cycling rate. The accumulators are recharged from a single or redundant supply line from the surface and the pilot valves are located in retrievable pods. When control line pressure is removed, the spring-loaded actuator dumps to sea through the pilot valves.

Although not as reliable as direct hydraulic control, the control pod is much simpler than other types of control system because in this case, fewer components are needed and there is no need for small orifices that can be susceptible to clogging. To minimize leakage, shear seal pilot valves are uses and the system is made relatively tolerant to contamination. Here in this type of control, control bundle costs are high and accumulators must be retrievable.

Sequential Hydraulic

In this type of control, pilot valves in the sub-sea controller are designed to a pre-set pressure level. Consequently, by varying the pressure in the control line, the pilot valves are operated in a predetermined sequence. As the practice is, only six to eight functions per signal line are accommodated. Further, as the length of the control line increases, fewer functions are available because it requires that the pressure step between sequences must be increased. Here, it is normal to use a reference pressure instead of a spring in the pilot valve to reduce the pressure interval over which the valve operates and to maximize the number of steps.

However, as the template trees require approximately twice as many functions as can be provided by one hydraulic control line, intermeshing two or three independent controllers to arrive at the desired sequence results in extremely complex hydraulic logic. Therefore, this control logic leads to an impractical approach. Further, since it is necessary to cause some valves to open and then close as the signal pressure increases, several additional pilot valves are necessary to implement the desired logic. Typically, sequential control systems have about three times as many pilot valves as there are actuators to be operated. To overcome these difficulties, it is possible to use a single hydraulic line to perform the control, reference and supply functions. This, however, reduces reliability because it requires pressure regulators and orifices to be at sub-sea. The only advantage of sequential control over discrete hydraulics is that it minimizes the cost of the hydraulic line bundle. There is no gain in response time, and reliability is considerably lower than with direct or discrete hydraulics.

Electro-Hydraulic Control

In electro-hydraulic control, a pair of wires is provided from the surface to a solenoid-operated valve for each valve to be actuated. The pilot valve is spring return and hence, requires power on continuous basis to maintain valve actuation. If dual pairs of lines are provided, bi-stable solenoid valves with a hydraulic fail-safe feature are used. This improves reliability but complicates the electrical connections.

The system is analogous to discrete hydraulics for simplicity, but it can be used over much longer distances and at a lower capital cost per unit distance. Compared to hydraulically-operated pilot valves, the solenoid valve requires much lower actuating forces, the valve is less tolerant to contamination and fouling and is therefore less fail-safe. Further, it requires the provision for position read back on all actuators which is done either by limit switches or by flow meters in the actuator line or by any other devices. One way of resolving these difficulties is to have a solenoid that operate a small pilot valve which then drives the main pilot valve hydraulically.

Multiplexed Electro-Hydraulic Control

The Multiplexed Electro-Hydraulic (MUX) control system is similar to the electro-hydraulic type. However, in this type, surface commands are processed electronically and transmitted in coded form through a single pair of wires to a sub-sea remote terminal unit. This type of control system is having the means to verify that the signal has been received correctly at sub-sea and the appropriate solenoid valve is actuated.

The primary advantage of the MUX system is its lower capital cost when distances are large. It can also be made significantly more reliable than the electro-hydraulic system by using dual solenoid bi-stable valves. In this case, the solenoid does not have to overcome a return spring in addition to frictional forces; and therefore the valve is more tolerant to contamination. This type of valve can be made fail-safe by using a reference hydraulic line to hold off a strong return spring. Upon loss of hydraulic pressure, the valve returns to its fail-safe position regardless of solenoid status. Nevertheless, solenoid valves have relatively low reliability. MUX requires sophisticated electronics and valve status indication and therefore these devises are comparatively less reliable and satisfactory. Further, as the electro/hydraulic interface (i.e. solenoid valves) is a weak point, these types of control system always need a back-up, usually a sequential hydraulic control system.

7.5 SUB-SEA FLOW LINES AND RISERS

The sub-sea flow lines and risers provide the connecting links for the flow path connection between the FPS and the template. The flow lines are connected to the template and riser base using divers and diver-assist connection systems. The lines are initiated at the production template and are laid toward the riser bases prior to installation of the risers. Second ends are laid down directly into the riser bases, with length discrepancies accommodated by a sliding "L" spool on the riser base. One important thing is that the length of these lines and positional tolerance of the bases relative to the template are critical. Export lines are initiated at the riser base of the chosen FPS and are laid toward the CALM buoy. Second-end connection is by a 180-degree sweep at the riser base for the buoy. Connection methods utilize come-alongs and Chinese fingers for manipulation of the pipe, together with swivel flanges and hydraulic bolt tensioners for make-up.

The riser bases are installed and anchor piled to the seafloor during the template installation stage. The risers are installed after the FPS is on station, using an appropriate Multi-Service Vessel (MSV) and divers. The risers are attached to the riser bases using diver made-up flange connections, and connect to the FPS by an emergency quick-disconnect hydraulic coupler.

Riser configuration is important. So it is important to undertake a static analysis of various riser configurations using the appropriate software. Riser configuration helps in ensuring that the risers are able to remain connected in the 100 year storm and the risers do not exceed minimum bending radii during all anticipated FPS motions. Riser configuration is specified by a host of parameters like buoy uplift in lbs, buoy height, length of catenaries, standoff distances and buoy-to-riser connector, the value of all these depends upon the sea-condition and also upon the type of FPS chosen,

In general, Flexible flow lines are selected over steel lines for the following reasons:

- Lower installed cost compared with double-insulated, rigid steel pipe.
- Ease of installation and recovery, and can be reused.
- Built-in thermal insulation and electrical heat-tracing.

Flexible production risers have a number of advantages. The flexible riser system can remain connected for both minor and major well work-over, permitting uninterrupted production operations while performing well work-over, if desired. The flexible riser system does not require the rig's mono-pool or draw-works, permitting their use for a second riser system to enable minor and major well work-over. The flexible riser system is comprised of a minimum number of components and permits a larger motion envelope for the semi-submersible than a rigid riser system would. Further, flexible risers have minimum deck loads associated with the riser system's static and dynamic loads.

Sub-Sea Production Systems: Considerations and Philosophies

After the decision has been made to utilize sub-sea production systems for field development, there are certain considerations like that of equipment configurations and field lay outs for the sub-sea production system, the methodologies adopted for installations of the components of sub-sea system, the sub-sea systems being wet or dry and like these many other general considerations that need to be understood in the field, reservoir, floating production and sub-sea system perspective. These understandings are important to work out a plan, configuration and lay-out for ensuring an optimum reservoir exploitation and production in cost-effective manner with high degree of operational feasibility and reduction in sea bed complexity.

Before we discuss these configuration, lay-out and general considerations, let's first understand the few basic differentials between a platforms based development and sub-sea based development:

Consideration	Fixed platform-based field development	Sub-sea based field development
Capital Cost	More expensive	Advantage (case specific)
Operating Cost	Advantage	More expensive
Time/Investment Profile	Longer period require greater investment	Advantage in term of time and investment
Production Profile	More recoverable reserves due to no longer field life	Earlier production—less recoverable due to higher operating costs
Water Depth	Advantage at < 1500 ft	Advantage at > 1500 ft
Net Pay	Requires more about 130 ft	100 ft
Field Area	Larger area—advantage minimum platform cost/well	Smaller area— advantage; fewer wells
Reservoir Depth	Medium depth reservoir	Shallow and deep reservoirs, advantage
Well Spacing	130-acre range	100-acre range
Development Time	Comparatively takes more time to develop	Advantage as it has minimum time for development

8.1 SUB-SEA PRODUCTION SYSTEM: EQUIPMENT CONFIGURATIONS AND FIELD LAYOUTS CONSIDERATIONS

The sub-sea equipment configurations and field layouts significantly influences the overall field development configuration. Depending on the specific requirements of the field development plan, the sub-sea production system may contain all or some of the following elements: (a) Sub-sea satellite wells (production and/or reservoir injection) and Christmas trees (b) Sub-sea templates and manifolds (c) Flow lines and d) Riser and riser bases. An optimal layout of these sub-sea production elements provides high degree of operational feasibility and reduces sea bed complexity. A reduction in the seabed layout complexity reduces the overall risk of damage from anchors or fishing activities.

Sub-sea production systems are generally configured and the field lay-out being done in one of the following ways:

8.1.1 Satellite Well Development configuration
8.1.2 Remote Template Well Development configuration
8.1.3 Template Well System Beneath a Semi-submersible configuration
8.1.4 Clustered Satellite Well Development configuration

Each of the configuration or lay-out listed above has its own set of advantages and disadvantages as discussed below. Understanding of each one of these in terms of its advantages or disadvantages is important so as to plan and configure the best possible lay-out plan for a given field development.

8.1.1 Satellite Well Development Configuration

Advantages

- This configuration has simple design.
- This configuration deploys minimum number of surface production facilities for field development.
- This configuration doesn't put any field development restrictions due to reservoir size and shape.
- This configuration has totally independent operation.
- In this configuration, minimum development time is possible because several rigs can be drilling simultaneously and immediate production can be taken after each well is completed.
- In this configuration, vertically drilled wells cost less to drill and maintain.
- Under this configuration, well service vessel is independent from production facility and therefore it gives minimum production loss during well servicing.
- This configuration provides the maximum vendor equipment flexibility.

Disadvantages

- This configuration requires a number of flow lines are which are expensive and increases the risk of damages by anchor and trawler gear.
- This configuration requires separate service rig for maintenance work.
- This configuration requires the fishing gear protection systems.
- This configuration has the maximum ocean floor coverage; hence it has the greatest risk for damage from non-production related events.
- This configuration requires that drill rig must reset anchor pattern after each well is drilled.

8.1.2 Remote Template Well Development Configuration

Advantage

- In this configuration, wells can be drilled with one anchor pattern.
- No well flow lines are required in this configuration if manifolding is done on template.
- This configuration has comparatively reduced number of surface facilities.
- For this configuration, protective devices for fishing gear and dropped objects are less expensive than satellite systems because of the common locations.
- This configuration has limited ocean floor coverage and hence, it has medium risk to damage from non-production related events.

Disadvantage

- This configuration requires a service rig for well servicing.
- Group pipelines and control lines to production facility.
- In this configuration, deviated wells are more expensive.
- This configuration requires protection for fishing gear.
- In this configuration, drilling starts after template installation and wells are drilled with one rig at a time. And hence, production can start only after all wells are drilled and completed.
- In this configuration, it is possible that all template production will be shut in during well maintenance.

8.1.3 Template Well System Beneath a Semi-Submersible

Advantage

- In this configuration, no flow lines are required.
- In this configuration, deployed floating production facility offers protection from fishing gear and vessel anchors in year-round production schedules.

- In this configuration, we can deploy common vessel for well remedial work and for production.
- Here, wells can be drilled with one anchor pattern
- This configuration has minimum ocean floor coverage and hence, it is least prone to damage from non-production related events.

Disadvantage

- In this configuration, drilling starts after template installation and wells are drilled with one rig per template.
- Under this configuration, deviated wells are more expensive.
- Under this configuration, we may have to go for the shut-in of the total production of a template during a work-over.
- This configuration has the maximum risk of damage due to dropped objects.
- This configuration has the greater risk of loss of surface facility due to well blowout.
- This configuration has one surface facility for each template.
- Under this configuration, production can begin only after all wells are drilled and completed.

8.1.4 Clustered Satellite Well Development Configuration

Advantage

- This configuration doesn't require any template.
- This configuration has the shorter lead time to start drilling.
- Semi-submersible deployed with this configuration is capable of doing work over of the wells.
- Semi-submersible overhead protects sub-sea equipment from fishing related damage and dropped anchor damages
- This configuration entails minimum lost production during well work-over.
- In this configuration, equipment covers minimum seabed area.
- This configuration has independent sub-sea structures. Hence, dropped objects impact single unit thereby minimizing lost production.
- This configuration as lower operating and maintenance costs than satellite well system configuration.

Disadvantage

- In this configuration, wells are drilled by one rig only at a time.
- Under this configuration, deviated well costs are high.
- This configuration requires short flow line jumpers to connect to riser manifold.
- In this configuration, sub-sea equipment is exposed to damages from supply boat dropped objects.

Besides these equipment configuration and field lay out consideration, we have two other very important considerations, as given below, which play very important roles in designing the configuration and lay out of the sub-sea equipment.

8.2 WET SUB-SEA PRODUCTION SYSTEM *VS*. DRY SUB-SEA PRODUCTION SYSTEM

The one very important consideration while configuring and designing layout is to decide whether we want a wet sub-sea system or dry sub-sea system.

In wet sub-sea installations, the components of the system are exposed to seawater and hydrostatic pressure. These components are installed or serviced either by running tools, or by remotely-operated vehicles, or by manned submersibles, or by divers. Most of the existing sub-sea production systems are of the wet type. Off-late, a lot of development and improvements have been carried out on wet system components to enhance their reliability. Earlier, wet sub-sea equipment consisted simply of slightly modified land components. But now, these wet system posses a philosophy of sub-sea repair and maintenance by retrieving a component or module to the surface and replacing it sub-sea with a new or surface-repaired unit. In relatively shallow water depth, wet sub-sea system is always preferred because of its simplicity and limited maintenance requirement.

Dry sub-sea systems benefit mainly from the advantage of direct human intervention for maintenance and repairs of its components in a one-atmospheric dry chamber. Apart from being able to use less expensive, more reliable and field proven land components, it has the advantage of the versatility and quality of service on complex systems through the trained hand (technicians). The major disadvantage is that the dry systems require a specialized service system.

8.3 DIVER INSTALLED, DIVER-ASSIST INSTALLATION, AND DIVER-LESS INSTALLATION

The other very important consideration while configuring and designing a lay out is to decide what kind of equipment installation interventions we wish to go for. We have basically three types of equipment installation interventions at the sub-sea:

- Diver Installed i.e. installation by divers
- Diver-assist Installation
- Diver-less Installation

Diver-installed systems require a considerable amount of diver intervention during installation to carry out such tasks as measurements, alignments, flange-ups, testing, operating valves, and connecting flow lines. Generally, diver-installed systems require divers to fit a sub-sea tubing head on a mud-line suspension system, to test tubing hanger seals, and to flange-up the manual tree connector to the tubing head. Normally, diver installed systems are used with wells drilled from a jack-up or bottom-supported rig.

Diver-assisted sub-sea systems reduce diver intervention work by performing several remote installations and testing tasks from the surface rig. This is accomplished by the use of specially designed equipment and running tools. The diver's role in this case is reduced only to monitoring, manual valve operation, and perhaps some testing and connection of the flow lines. These systems are more expensive than diver-installed systems. These diver-assisted sub-sea systems are normally used with wells drilled from floating drilling rigs.

Diver-less sub-sea systems are designed to enable total remote installation, testing and operation of the sub-sea system. Divers are not required, as in this case, even the flow lines are capable of being installed remotely. The system relies on atmospheric diving suits, JIM, WASP, etc. The system relies on ROV for inspection and monitoring and also for performing minor in-situ maintenance tasks. Here, the primary maintenance philosophy is to recover the equipment to surface for repair/replacement. To minimize equipment recovered, the equipment is designed into retrievable modules, control pods, seal plates, etc.

For a water depth of 60 to 120 meters (200 to 400 feet), diver-installed and diver-assisted installation systems works quite well as the water depths are well within the range of effective diver intervention. For water depth beyond 180 meters (600 feet) where diver intervention deemed economically infeasible and difficult, diver-less sub-sea systems works well.

8.4 SUB-SEA EQUIPMENT AND SYSTEM: DESIGN, MAINTENANCE AND OPERATION PHILOSOPHY/STATEMENT

In any sub-sea production system, every equipment or component has a design, maintenance and operation philosophy statement, which we sometimes refer as guideline policies. Developing a philosophy statement (and on the similar lines, the requirement and goal statement) for each and every component of the sub-sea system is quite an exhaustive exercise. Nevertheless, this exhaustive statements or guidelines policies specific to each equipment or component, either individually or in a group, do exist. And hence, an understanding of this is required and expected.

For example, here, I have discussed the design philosophy with respect to the satellite tree; maintenance philosophy with respect to down-hole maintenance and an operation philosophy with respect to normal production operation. I would like the student to develop as much understanding of the similar philosophies or guidelines policies or statement for each and every component/equipment of the system as possible through referring design books, or by going through specific journals/magazines or through their interaction with experts. It is important.

By term "philosophy", I mean here, a statement which defines the basis overall principles of the system. By term "requirement", I mean here, a set of condition which the sub-sea systems *must* meet. "Goals" are generally less specific statements which *should* be met.

8.4.1 Design Philosophy Statement: Satellite Tree

- Reliability through simplicity.
- Maximum use of field-proven equipment and techniques
- Ease and speed of installation and maintenance, with minimum sensitivity to weather conditions.
- Total interchangeability of components that perform the same function.
- Minimum number of seals made up sub-sea during installation, production or maintenance operations. No seal surfaces to remain unprotected sub-sea for extended periods.
- Minimum numbers of electrical make and break connectors sub-sea.
- Independence of production path should always be considered in order to avoid total shutdown from one fault.
- Planned diver intervention to be used where a significant reduction in complexity or cost is evident. All equipment configurations should be designed for safe access either by divers or submersibles.

- Ability to troubleshoot an equipment malfunction from the production platform in preparation for maintenance.
- Safety-conscious approach to all aspects of design and maintenance
- Guideline drilling system to be used
- Equipment design life of 20 years. Equipment to be inspected and refurbished prior to reinstallation for a second application

8.4.2 Maintenance Philosophy Statement: Down-Hole Maintenance

The intent of the maintenance philosophy is to maximize equipment simplicity and reliability. It implies that sub-sea equipment to be used should be the least complex, which in-turn should contribute to the least expensive maintenance and operating costs. The general maintenance philosophy statements for sub-sea equipment installed on the ocean floor are as follows:

- Maximize simplicity of equipment, to minimize maintenance requirements and to simplify maintenance tasks required.
- Maximize the use of divers, using diver-installed or diver-assist sub-sea equipment types
- Maximize diver intervention for in situ maintenance and inspections

Down-Hole Maintenance Philosophy (major work-over and minor well-servicing)

Down-hole maintenance philosophy with respect to major well work-over requires that the effort must be placed on the well completion programme and on the selection of completion equipment so as to avoid any such major work-over interventions and to ensure maximum reliability of the reservoir's productivity. Major well work-over are normally not carried out in a sub-sea well in a marginal field owing to the expenses involved and owing to its non-economic feasibility considerations. Further, it is also not feasible economically if we are exploiting a field through sub-sea wells and floating production systems because the major work-over exercise here requires a work-over vessel whose call-out costs are very high.

Down-hole maintenance philosophy with respect to minor well servicing (either through a wire-line or through TFL, through tubing flow line) requires the selection of a system for simplicity, while minimizing the associated added complexity to the sub-sea hardware. The maintenance requirement for minor well servicing should be such that it should provide simplest well completion for reliability considerations, it shouldn't impact the complexity of other hardware (tree, flow line) and should be the least expensive besides being the system most commonly used.

8.4.3 Operating Philosophy Statement

Normal Production Operation Philosophy

- To enable environmentally safe production operations on continuous basis
- Remote individual well control
- Flow line circulation/flushing
- Remote well and flow valve testing capability
- Independent well production—one failure shouldn't shut in all production
- Ability to monitor pressure (annulus and production)
- Remote well/annulus kill capabilities
- Ability to bleed-down the annulus
- Down-Hole Safety Valve (DHSV)
- Remote testing capability for DHSV and master valves
- Automatic emergency shutdown capability of wells if abnormality is detected
- All remote-operated equipment to be capable of operation by divers as a backup.

Sub-Sea Production Systems: Selection Criteria for Different Floating Production Systems

When we talk of choosing sub-sea production system and its components for the three types of floating production systems (Tanker, Barge and Semi-submersible), we mean how the sub-sea wells, templates, manifolds, well completions and well control systems are chosen for a given floating production system. The pertinent issue here is that whatever be the criteria, the choice must ensure better compatibility, improved operational flexibility, least complexity, reduced cost and should present a comprehensive whole of the given FPS and chosen sub-sea production system.

9.1 BARGE-BASED SYSTEM: SUB-SEA PRODUCTION SYSTEM OPTIONS

9.1.1 Sub-Sea Wells, Templates and Manifold

The sub-sea production systems utilizing the barge-based floating production systems traditionally have been sub-sea satellite wells, without any manifold. The reason being that the barge-based systems are normally used for small field developments of one to four wells or for single-well extended test systems. This small number of sub-sea wells does not warrant the use of sub-sea manifolding or template systems.

9.1.2 Sub-Sea Well Completion

For Barge-based floating production systems, sub-sea satellite well completions are selected. This type of completion system is relatively simple and inexpensive. The tree assembly for the chosen completion is a diver-installed, wet satellite tree, with tree sizes with specified working pressure depending upon the given production and reservoir plan. This tree assembly comprises of tree piping and X-mas tree with a number of individual valves and components.

It is better to understand the well-completion in terms of completion terminology and specification. Figures given here and also ahead in this chapter are indicative only just to aid in comprehending the system and its components. For example, here in the instant case of field development, the 5000 psi working pressure (wp) tree is assembled from 3–1/8" and 2–1/16" ID API valves and components. 3–1/8" Production bore equipped with fail-safe close hydraulic operated actuators and manual override is used for lower master manual valve, for upper master, swab calve and wing valve. 2–1/16" Annulus access equipped with fail-safe close hydraulic operator with manual override is used for side outlet access provided on the tubing head and also with manual wing. X-mas Tree piping is a 3–1/8" ID, 5000 psi wp steel piping. Production piping terminates on tree assembly with 3–1/8" 5000 psi wp manual production flow line valve. Flow line connection is diver-assisted. Annulus is tied into production piping so as to permit remote annulus bleed-down or well kill by circulation via the flow line and permanently installed wire-line riser (wells beneath barge only). The X-mas tree also has a manual pig-entry system to permit production flow line pigging.

9.1.3 Sub-Sea Well Control System

The preferred sub-sea control system for the barge-based FPS option is "direct hydraulic with no sub-sea data acquisition system". The sub-sea control system for the barge-based FPS option consists of all the equipment required to operate the tree valves and the Surface-Controlled Sub-sea Safety Valves (SCSSV).

Sub-sea well control system, here in barge case, consists of the following:
- A surface Hydraulic Power Unit (HPU) mounted on the main deck of the barge, generating hydraulic pressures to operate hydraulic actuators on the tree and for SCSSV. The HPU is interconnected with the FPS ESD system to cause automatic shutdown of the HPU in the event of an emergency, thereby effecting the shutdowns at sub-sea production systems.
- A self-standing production control panel for each well placed preferably in the barge control room (it can be outside the control room too with remote control and monitoring features added to the panel so as to operate from control room).
- A hydraulic umbilical for each well comprising of a number of reinforced hoses. Umbilical terminating at tree-end has a diver-replaceable flexible jumper hose assembly that helps divers to make up sub-sea connection with the tree quite easily.
- A tree-mounted equipment consisting of the tree half of the umbilical connector, ½-inch OD stainless steel lines from the connector to each

valve actuator (each line has an isolation valve) and isolation valves to connect the spare line to each other line.

- The hydraulic piping on the tree and to the SCSSV's.

As far as data acquisitions are concerned, no capability is provided to monitor sub-sea well parameters except those for monitoring the annulus. The annulus monitoring line is connected between the annulus manual master valve and the automatic annulus wing valve, and runs to the surface in the control umbilical to a pressure transducer installed on the control panel.

9.1.4 Sub-Sea Production System Maintenance

The chosen wet tree system utilizes divers for installation and maintenance. Down-hole servicing is carried out by surface operations, either from the barge for the wells located beneath the barge, or from a mobile service rig for the other satellite wells.

9.2 TANKER-BASED SYSTEM: SUB-SEA PRODUCTION SYSTEM OPTIONS

9.2.1 Sub-Sea Wells, Templates and Manifold

For tanker-based system, a number of sub-sea systems options like Sub-sea template option, Sub-sea manifold and satellite wells option, Total development by sub-sea satellite wells options are available. However, it is always advantageous to go for sub-sea satellite wells options. Some of the advantages with this option are as follows:

- It has least impact to development schedules.
- It has minimum costs.
- This sub-sea satellite wells option has simplest sub-sea equipment configuration. Though the sub-sea template and sub-sea satellite wells with manifold options reduce the number of flow lines in the riser system by sub-sea manifolding, these options, however, increase the complexity of the sub-sea systems by placing more equipment (i.e. the manifolds) at sub-sea. The increased amount of equipment at sub-sea may result in increased installation and maintenance costs.
- Wells positioned to optimize recoverable reserves.
- Wells are drilled vertically to minimize drilling costs.
- Maximizes production availability as sub-sea equipment failure or well work-over result in only one well being shut in, therefore having minimum lost production due to maintenance or equipment failures.

9.2.2 Sub-Sea Well Completion

The well completion in tanker FPS case differs from that of the barge FPS case. In tanker based system, sub-sea well completion is a 3–1/8" × 2–1/16" well completion with vertical access through the tree and tubing hanger for both annulus and production services. The tree assembly is a wet diver-assist tree assembly. The tree is capable of remote installation, testing, and tieback to a surface rig for well maintenance work. Divers are used to make up the flow line and control umbilical connections and also for observation and maintenance as required.

To understand the tree assembly specification (figures are indicative only), in the instant case, the tree assembly is a wet, 3–1/8" × 2–1/16" dual mono-block tree rated for 5000 psi wp with integral swab valves and tie-back mandrel to facilitate rig tie-back. The tree has a hydraulic wellhead connector (16–3/4" 5000 psi), guide frame assembly (4-funnel, 6-foot radius guide frame) and remote lock, unlock and test capabilities. The 3–1/8" Production bore of the tree is equipped with fail-safe close hydraulic operated actuators and manual override is used for lower master valve, for upper master valve, swab valve and wing valve. The 2–1/16" annulus of the tree is used for side outlet access provided on the tubing head and is equipped with manual wing valve and fail-safe close hydraulic operator with manual override. Wing Valves of tree assembly consists of 3–1/8" and 2–1/16" target cross-wing valves equipped with fail-safe close hydraulic operators and manual override. Crossover of tree assembly is 3–1/8" valve equipped with fail-safe open hydraulic operator, and manual override. Tree Cap of tree assembly is 5000 psi, 13–5/8" hydraulic connector with integral guide frame assembly. Flow line connections are diver-assisted and are arranged to facilitate round-trip pigging via production/annulus crossover.

9.2.3 Sub-Sea Well Control System

The control system for the tanker-based production system differs substantially from the system described for the barge system. Though sub-sea hydraulic parts of the chosen well-control system remain more or less the same in both the cases, they differ in each other with respect to the surface based hydraulics and other control systems. Maintenance-wise, sub-sea well control systems are easy to maintain at tanker based system as compared to barge based system simply because the complicated parts of the control system are readily accessible from the tanker.

In the tanker based FPS case, direct hydraulic with totally independent data acquisition system are used. Here, the sub-sea well control system consists of the following:

- A Hydraulic Power Unit (HPU) located on the rotating part (swivel) of the SPM (single point mooring) and bulk hydraulic fluid storage on the tanker. This HPU can also be placed on the tanker itself, at the expense of additional hydraulic lines passing through the yoke. However, placing on tanker option is slightly less reliable than the option when HPU is placed on swivel.

- A multi-pass swivel at SPM having redundant hydraulic lines in addition to redundant electrical power and signal lines. The signal lines carry frequency shift keyed digital two-way communication from the control room panels to the multiplex controllers attached to the riser.

- Multiplex electro-hydraulic controllers attached to the risers below the swivel but above sea level. This type of controller is necessary to minimize the complexity of the swivel which would otherwise require a large number of hydraulic passages. These arrangements provide direct hydraulic control to the sub-sea valves since there is no control valves required sub-sea.

- A self-standing production control panel for each well placed preferably in the barge control room (it can be outside the control room too with remote control and monitoring features added to the panel so as to operate from control room).

- Electro hydraulic umbilical to each of the six wells. Outputs from the controllers are fed to individual umbilical for each well. The umbilical for each well are similar in construction to that of the barge-based system, except that in this tanker case, it contains one additional ¼-inch hose and four pairs of electrical wires to provide power and data communication to the sub-sea electronics for the instrumentation.

- A well-mounted equipment.

In the tanker-based system, sub-sea well control system also has provisions for data-acquisitions. For this, sub-sea instrumentations are provided at the sea-bed which gives us data for wellhead fluid pressure and temperature, annulus pressure and down-hole pressure and temperature. Outputs from these instruments are fed to a Remote Terminal Unit (RTU) which contains a microprocessor to process the data. The data is logged on the surface automatically by a computer, and alarms are generated as dictated by operational requirements. Interconnection between the RTU and the umbilical is effected with a wet mateable sub-sea electrical connector.

9.2.4 Sub-Sea Production System Maintenance

The wet tree system can be remotely installed and tested from a floating drilling or completion rig. This system can also be remotely tied back to a surface rig for wirelining, major well work-over or other maintenance tasks. Divers are used for observation and minor in situ maintenance tasks.

9.3 SEMI-SUBMERSIBLE-BASED SYSTEM: SUB-SEA PRODUCTION SYSTEM OPTIONS

A number of sub-sea production system options are available for being considered with a semi-submersible-based floating production systems:

- Total development using satellite wells
- Sub-sea manifold and satellite wells
- Sub-sea template beneath FPS
- Clustered satellite wells beneath FPS

Each one of these are having their own set of advantages and disadvantages and choice of any depends upon a host of factors like reservoir conditions, production requirements, and semi-submersible facilities considerations as well as considerations that are inherent to semi-submersibles.

For example, when a semi-submersible vessel is selected as the FPS and reservoir considerations permit deviated wells to be drilled from a common seabed location, sub-sea templates or satellite well clusters should be used. It is always better to place the sub-sea wells immediately beneath the semi-submersible vessel, thereby getting benefited from the features that are inherent to a semi-submersible FPS. This type of well placement enables the rig to perform well maintenance work (like that of wire-line and minor and major well work-over). As the semi-submersible performs the dual role of a process and well maintenance vessel, the semi-submersible contributes to lower field maintenance and operating costs when compared to other field developments requiring mobile services rigs and vessels. Further, the diving spread of the rig can also be used for diver tasks thereby avoiding the need to call out a separate diving spread or work vessel. Hence, in view of these, the two preferred sub-sea development options considered for the semi-submersible FPS are: Template development and Clustered satellite well development.

In both these sub-sea system options, wells are located directly beneath the semi-submersible rig. Further, for both the case, flexible risers are used over rigid tubular risers. Flexible risers do not need a mono-pool, thereby freeing the mono-pool area for well work-over systems and hence avoiding the lost production associated well remedial work and the necessity for production riser disconnects. Flexible risers also do not require tensioning systems. These flexible risers also don't require storage onboard because they are never pulled and hence this permits more deck space and load-carrying capability for the process equipment.

Further, if we have to choose one out of the above discussed two options, it is always preferable to use sub-sea template system for the following reasons:

- Capital cost is less expensive
- Fewer flexible dynamic risers, which implies lower maintenance costs
- Fewer risers for rig to handle, thus making it:
  - is easier to accommodate (lay out) all risers on one side;
  - enables greater deck load capability;
  - requires smaller emergency quick-disconnect system

Further, it is also preferred to go for the Template manifolding as it decreases the number of flexible risers.

To conclude, the sub-sea template system and template manifolding is the best suited option for semi-submersible floating production system option.

Offshore Crude Storage, Heating, Loading and Transportation

Offshore Crude Storage, Heating, Loading and Transportation are important consideration while planning for the development of any new oil field through floating production system. It requires comprehensive deliberations over whether to lay a pipeline, or use the existing pipeline of nearby fields or use an offshore storage/tanker loading system. Economic feasibility is an important consideration here.

It is prudent here to here to have good understanding of Tankers and Super-tankers along with the types and tonnage and weight measurements principles involved. Some basic fundamentals have been discussed at the end of this chapter.

In this chapter, we shall discuss the following:

10.1 Offshore Storage Facilities
10.2 Crude Offloading System
10.3 Oil Storage Heating System
10.4 Tankers and Supertankers; Tonnage and Weight measurements

10.1 OFFSHORE STORAGE FACILITIES

10.1.1 Offshore Storage Facilities Categories

Offshore storage facilities can be classified into five categories. These are:

- Floating Captive Tankers
  e.g. Nido field, Philippines; Fulmar field, North Sea, U.K.
- Floating Purpose-built Storage Structures
  e.g. SPAR, Brent field
- Storage Facilities Incorporated Within a Production Platform or Tanker
  e.g. Brent B&D concrete gravity platforms; Maureen steel gravity platform, in the North Sea.
- Submerged Purpose-built Storage Structures
  e.g. Ekofish Tank, North Sea
- Barges
  e.g. Conoco's Kepiting field, Indonesia

The choice of types depends upon where we are going to use it i.e. whether we are using it in more harsh sea environment (like North Sea) or in less harsher sea environment (like Offshore India) and whether the economics is justifying or not the chosen type. Further, as submerged systems are often specifically designed to operate in given circumstances of particular water depths, it limits its flexibility in redeployment, unless an existing one is utilized. Sometimes, semi-submersibles are used for storage. However, such a vessel rarely has sufficient storage capacity, thus the crude must be directly produced/ transferred into the shuttle tanker.

The use of a barge or tanker is often the most practical means of storage. Certain advantages of the floating captive barge or tanker are as follows:

- It can be procured and deployed quickly, and thus is very attractive for early production of the field.
- It can be relocated to other fields.
- It is the least expensive type of storage being considered at this time.

10.1.2 Sea-State Considerations

While designing offshore storage facilities, adequate and in-depth considerations must be given to sea environment i.e. "sea-state" in which it is going to be deployed and also the climatic conditions like extent of day light available.

The sea-state affects whether the shuttle tanker can moor either to the single-point loading buoy or to the tanker-or barge-based FPS. Also, the sea state influences the choice of loading modes; tandem or alongside mode. In higher sea states, the shuttle tanker may have to discontinue loading. A contingency of one to two days should be allowed for weather and scheduling delay of the shuttle tanker or barge.

Depending on the area of operation, the transfers may have to take place in darkness. In the North Sea, for instance, the daylight hours can be reduced to six in any day. It is therefore essential that adequate lighting be installed in the critical areas, such as the forecastle area. This would involve floodlighting, searchlights, etc. It is emphasized that good visibility is of the utmost importance.

10.1.3 Shuttle Vessel Requirements

Working out an adequate shuttle vessel requirement is very important as it involves cost and time in big way. The number and type of shuttle vessel is determined by:

- Throughput of oil (production)
- Distance from port of discharge

- Speed of shuttle
- Carrying capacity of shuttle
- Efficiency of port equipment

In general, each shuttle is equipped with:

- A mooring winch to receive and secure the mooring hawser.
- A hose-handling winch, gantry, or davit to handle the hose and position it for connection to the offloading facility
- A hose coupling system to clamp the hose and secure it into position
- Power supply to the motors of the various units
- Control cabin to house controls
- Adequate oversight lights
- Communication system
- Control system to control the flow of crude

Vessel Selection Criteria for Shuttle Application

The vessel selections are normally based on the following criteria:

- Volume to be transported; if the vessel is too large, it will result in un-economic operation.
- Ability to handle the crude, i.e. flow rates, metering, pumping, surges, storage margins etc.
- Vessel speed and fuel consumption
- Tank arrangement
- Condition of vessel, its cost and age
- The mooring arrangements shall be adequate for the specific vessel
- Hose connecting:
  - The area shall have good accessibility and visibility
  - Valves should have clear position indicators
  - The hose-connecting unit should be simple and reliable, with backup facilities and rope load tension mounting system
  - Break spool coupling should be in hose string
- The power unit for installation onboard is normally hydraulic. This unit may be driven by an electric motor or diesel engine. It should be located in a safe area. The unit should be large enough to serve all consumers.
- Lighting

10.1.4 Offshore Storage Facility Considerations for Different FPS Options

Let's get ourselves appraised of the various facets and considerations of the offshore storage facilities with respect to the field under consideration (refer case study) under different FPS scenario. Figure reflected here are somewhat close approximations indicated only just for the sake of understanding.

Barge-Based FPSO Concept

The FPSO barge is required to store production for a period of time which is equal to the round-trip time of the shuttle. For the field under consideration here, it works out to 4.5 days. Accordingly, the production to be stored works out to be 45,000 to 50,000 barrels, giving due consideration to maximum heel and fill percentage. Against this requirement, a dedicated shuttle is normally preferred. It shall have a minimum storage of 50,000 barrels of oil. An 8000-DWT cargo oil barge will be sufficient. The proposed loading rate is 3000 bph (476 m³/h). Loading time is about 17 hours.

Tanker-Based FPSO Concept

The FPSO tanker is required to store production for a period of time which is equal to the round-trip time of the shuttle tanker. For the field under consideration here, it works out to 5.5 days. Accordingly, the production to be stored works out to be 110,000 to 122,000 barrels, giving due consideration to maximum heel and fill percentage. A 130,000 DWT tanker is preferred to be used as the FPSO tanker for economic reasons and better motion characteristics. The maximum storage available is 700,000 barrels of oil. One shuttle tanker will be required. It shall have a minimum storage capacity of 110,000 barrels of oil, or approximately 17,000 DWT. However, for economic reasons, a 50,000-DWT tanker is preferred, as the unit transportation cost will be the lowest. Since the shuttle is over-sized, it should not be dedicated to serve the field alone. It should also serve other fields in the vicinity. A 50,000-DWT shuttle tanker can transport approximately 375,000 barrels of oil. Based on a loading time of one day, the loading rate is worked out to be 16,000 bbl/hr.

Semi-Submersible-Based FPS Concept

Semi-submersible based FPS normally has a minimal storage already onboard and the production is normally pumped directly to the shuttle tanker. The minimum shuttle tanker size is approximately 17,000 DWT to contain 5.5 days of production. The number of shuttles required is two, assuming one being loaded, while the other is en-route to shore and back. However, for economic reasons, two 50,000-DWT shuttle tankers shall be used. They will not be dedicated to serve the field alone. During shuttle tanker changeover, the production has to be produced temporarily to the onboard tanks. A storage capacity for 8 hours' production is normally preferred that works out to be 8000 barrels of oil storage with a loading rate of 1000 bph.

10.2 CRUDE OFFLOADING SYSTEM

Crude Offloading System comprises of the following:

10.2.1 Pumping System/Crude Transfer Pumps

10.2.2 Stripping System

10.2.1 Pumping Systems

For crude offloading, conventional Centrifugal Cargo Pumps are commonly used in tankers. There are basically three types of centrifugal cargo pumps:

(a) Horizontal split-case cargo pumps, preferably with an external bearing arrangement
(b) Vertical split-case cargo pumps
(c) Barrel-type cargo pump

As space is at a premium, especially on a tanker, by far the majority of cargo pump arrangements are vertical, thereby, reducing the size of pump room required. Now a day, the preferred arrangement of vertical-type cargo pumps is the vertical overhung, barrel-type pump unit. From the point of view of the ease of onboard maintenance, a new type of pump known as vertical overhung impeller cargo pump, referred to as the barrel-type cargo pump, has been developed. The overhung impeller-type cargo pump, referred to in future as the barrel type, normally has the same basic hydraulic design as an equivalent-sized, conventional vertical/horizontal split-case cargo pump. Indeed, it is often possible to use the same design of impeller. The suction/discharge of the barrel type pump can be either of the straight-through arrangement or at a 90 degree suction/discharge angle. Both connections are in the main bottom half casing. The bearing arrangement of this type of pump is completely different from the type previously described, as bearings are placed above the impeller. The top bearing consisting of a double bearing acts as a radial-load bearing. The difference between the top and bottom units is extremely important in ensuring that the complete rotating unit is stiff, preventing any tendency for the impeller end to wipe.

Barrel-type cargo pumps are normally fitted with mechanical seals, and it is also possible to fit a split seal. The make and type of mechanical seal fitted are normally as per the owner's preference. With this design of cargo pump, the impeller can be situated lower in the cargo pump room than the impeller of a vertical, conventional split-case type and thus improving the suction conditions to the impeller. The barrel-type pump is designed with special attention to easy onboard maintenance; the most important aspects being:

- Removal of the complete rotating element without disturbing discharge pipe work.
- Removal of bearings and shaft seal without removing the rotating element.
- Access to mechanical seals through apertures in pump housing
- Provision for split seals to be fitted, if required.
- Provision of well-positioned lifting points to enable complete and easy removal of the rotating element.

The materials of construction for both types of cargo pump are selected in accordance with the type of products that the pump will handle.

10.2.2 Stripping Systems

The primary requirement of the loading system is that the cargo pumps have to be capable of stripping the tanks free of cargo to the specific level. Vertical centrifugal pumps will be able to pump the cargo down to near the centerline of the impeller. To go lower than this, other low-suction screw pumps or reciprocating piston-driven pumps will have to be incorporated. Stripping systems are required for the barge-based and tanker-based FPSO systems. They are normally not required for the semi-submersible-based FPS.

On tanker based FPS or Barge based FPS, we have to see whether the tanker or Barge have sufficient pumping capacity to offload its crude at the required rate, say for example, of 16,000 bph and accordingly we should see whether the cargo pumps and stripping systems already available on the tanker requires any modifications or not. On tanker based FPS, cargo pump system, for the field under study, consist of three steam turbine-driven vertical pumps of 8000-bph capacity each, whereas it is two cargo pumps (say of capacity 3000 bph) and two stripping pumps for Barge based system. The bottom-line is that the total pump capacity on board should be 150% of the required loading rate. The stripping system normally consists of independent reciprocating steam piston-driven stripping pumps. Number of cargo pumps and stripping pumps depends upon the offloading requirement of the fields under consideration.

10.3 OIL STORAGE HEATING SYSTEM

The temperature of the crude has to be maintained above its cloud point so that wax in the crude should not crystallize. This crystallization problem arises either in high pour point crude or in cold climatic conditions, or both. Further, in spite of best of Temperature maintenance efforts, some heat losses from the storage system are always there. Accordingly, a heat loss calculation, considering the climatic conditions, is normally done to determine the heating capacity. Some spare heating capacity also gets incorporated to deal with unforeseen circumstances.

10.3.1 Heating Methods

There are two methods of heating the crude in storage:

- Internal coil heating method which is widely used in the tankers and land-based storage tanks.
- External heating, in which crude is circulated through an external heat exchanger or fired heater by means of a circulating pump.

Internal Coil Heating

Many refineries and petrochemical plants use internal heating coils for heating crude oil, raw materials, products, and fuel oil. Though steam, normally of pressure from 50 to 150 psig, is the most common source of heat, electricity, "Dowtherm" and other heating media are also used when high temperature is required,. While considering the heating medium for crude heating, it is important to ensure that the skin temperature of the heating coil does not exceed the cracking temperature of the crude. It is difficult to estimate the cracking temperature of unidentified crude; however, temperatures above 360°C should be avoided. A steam heating system operates at temperatures much lower than this.

There are many different types of coil commonly used for tank heating. It is important that a coil design should offer good tank bottom coverage. Proper mixing of cell contents will eliminate local hot spots in the cell and will prevent sludge and wax from building up at the bottom of the cell or in a dead spot, as these phenomenons reduce the heat transfer from the coil.

External Heating

External heating uses a circulating pump and a heat exchanger for heating the crude outside the storage tank. The major advantage of external heating over internal coil heating is that the heating device and circulating pumps are located outside the storage cell, thus making maintenance relatively easy.

Pumps for circulation can either be large cargo transfer pumps or separator pumps. Heating media for heat exchanger can be steam, hot fluid, electricity, or exhaust gas from gas turbines or diesel engines. Use of fired heaters, normal process heaters, or salt bath heaters may also be justified, depending on the safety regulations and economics.

10.3.2 Heating Media

The heating media to be used for internal coil heating should be safe, if a leak or failure of the system occur. The following are the commonly used heating media for heating coils and heat exchangers.

Steam

Steam is commonly used in refineries and petrochemical plants. It is also widely used for heating waxy, high pour point crude in tankers. Steam's latent heat of condensation, 864.5 Btu/lb for 150 psia saturated steam, together with the large temperature differential available, makes it attractive to use as a heating medium since this will make the surface area of the heating coil smaller.

Hot Liquids

In cold climates, ethylene glycol (EG)-water mixture is commonly used in small-scale, closed cooling systems, and for heating of internal combustion engines, compressors, and pumps. Since boilers are installed on board the barge and tanker, the use of glycol (EG)-water mixture is not considered economical.

Hot Water

Hot water can also be used as a heating medium. It has similar advantages and disadvantages as steam; however, considerably larger coil surface will be needed because of its lower heat capacity.

Electricity

Although electricity is not a heating medium in the true sense, it is discussed here as a potential source of heat. Crude can be heated directly by immersion electric heating coils/tubes. However, if an immersion electric heater is to be used to heat the crude in the cell, it will not be possible to achieve uniform heating. Furthermore, there will be an overheating hazard, should the heating elements not be immersed in the crude.

10.3.3 Normal Industry Practice: Heating Methods and Heating Media Options

An internal heating system, using steam heating coils, is normally used for both the barge-and tanker-based FPSO systems. Two reasons can be attributed to this:

- Should the crude congeal in the tanks due to an emergency or due to the breakdown of the boiler, the internal coils are able to raise the temperature of the crude once the steam supply is resumed. On the other hand, for the external heating system, it will be difficult to raise the temperature of the crude, if the crude congeals in the tanks.
- The external heating system uses the heated, high pour point crude as the circulation medium; to prevent crude from congealing inside the pipes and pumps, heat-tracing will be required. This will be costly.

For the semi-submersible FPS, heat loss and the possible solidification of stored crude is not a problem for an 8-hour storage; therefore, storage heating is considered unnecessary. In case long-term storage becomes necessary at semi-submersible FPS, a pour-point depressant will be added.

Before we end this chapter, it is prudent to have some understanding of the basic fundamentals of Tankers and Supertankers along with the associated Tonnage and Weight Measurement principles.

10.4 TANKERS AND SUPERTANKERS

10.4.1 Tanker

A tanker is a ship designed to transport liquids in bulk. Tankers range in size and capacity from several hundred tons, which includes vessels for servicing small harbors and coastal settlements, to several hundred thousand tons, for long-range haulage. Further, tankers are built and classified as per the type of product it carries like "chemical tanker", "oil tanker", "liquefied petroleum gas (LPG)" tanker, "liquefied natural gas (LNG) tanker" etc as different products require different handling and transport.

Tanker Size

Tankers used for carrying liquid fuels are classified according to their capacity. Though in earlier days, "afra" system (*average freight rate assessment*) used to classify tanker sizes, now-a-days, tanker sizes are classified as per "dead weight tonnage (dwt)." Accordingly, Tankers are classified as below:

- 10,000–24,999 dwt: General Purpose tanker
- 25,000–44,999 dwt: Medium Range tanker
- 45,000–79,999 dwt: LR1 (Large Range 1)
- 80,000–159,999 dwt: LR2 (Large Range 2)
- 160,000–319,999 dwt: VLCC (Very Large Crude Carrier)
- 320,000–549,999 dwt: ULCC (Ultra Large Crude Carrier)

Single Hulled Tankers and Double Hulled Tankers

Tankers are also classifieds as per the "Hull": Single hulled tankers and Double hulled tankers. In "single-hulled" tankers, the hull is also the wall of the oil tanks, and any breach results in an oil spill. In "double-hulled" tankers, a space is available between the hull and the storage tanks to reduce the risk of a spill if the outer hull is breached. This space is used to carry water ballast when the ship is not carrying an oil cargo. In practice the addition of an extra hull should prevent such a ship from suffering a catastrophic breach of the hull. A double-hull tanker is generally safer than a single-hull in a grounding incident, especially when the shore is not very rocky. However, double hulls are not a complete solution, and they are at greater risk of explosion if petroleum vapor collects in the space between the hulls.

10.4.2 Supertankers

"Supertanker" stands for world's largest tanker ships having deadweight tonnage (dwt) of above 250,000 and capable of transporting around two or three million barrels of oil. "Supertanker" is an unofficial term, and in the shipping industry, Very-Large Crude Carriers (VLCC) and Ultra-Large Crude Carriers (ULCC) are now-a-days commonly designated as supertankers, though in early days, tankers having dwt quite lesser were known as supertankers.

Due to their size and mass, supertankers have very poor maneuverability. The stopping distance of a supertanker is typically measured in miles. When operating close to the shoreline they are vulnerable to running aground, largely because at slow speed it is impossible to control their movements. When this happens, there are chances of oil spills and accidents.

10.4.3 Tonnage Measurements: Concepts and Principles

"Tonnage" is a measure of the size or cargo capacity of a ship. The term derives from the taxation paid on *tuns* of wine, and was later used in reference to the weight of a ship's cargo; however, in modern maritime usage, "tonnage" specifically refers to a calculation of the volume or cargo volume of a ship. The term is still sometimes incorrectly used to refer to the weight of a loaded or empty vessel. However, measurements of tonnage are not so straightforward as it seems to be. One important thing to understand is that "Tonnage" and "Ton", are two totally different concept. "Tonnage" refers to the unit of a ship's volume in measurement for registration and "Ton" refers to the unit of weight.

Definitions of Deadweight and Lightweight for Tankers

Now let's understand what "deadweight" and "lightweight" means.

"Deadweight" is the difference in metric tones between the displacement of a ship in water of a specific gravity of 1.025 at the load waterline corresponding to the assigned summer freeboard and the lightweight of the ship. (taken From "Safety of Lift at Sea 1974":). Deadweight (often abbreviated as DWT for deadweight tones) is the displacement at any loaded condition minus the lightship weight. It includes the crew, passengers, cargo, fuel, water, and stores.

"Lightweight" is the displacement of a ship in metric tones without cargo, fuel, lubricating oil, ballast water, fresh water, and feed water in tanks, consumable stores together with passengers and crew and their effects. (taken from "Safety of Lift at Sea 1974":)

10.4.4 Tonnage Measurements Principals

We have different types of tonnage measurement principles in vogue. Knowing measurement system and its standardization is important because a ship's registration fee, harbour dues, safety and manning rules etc, are based on its measurement whether it is gross tonnage, GT, or net tonnage, NT. Tonnage measurements are governed by an IMO Convention (International Convention on Tonnage Measurement of Ships, 1969 (London-Rules)), which applies to all ships built after July 1982.

Gross Register Tonnage (GRT)

GRT represents the total internal volume of a vessel, with some exemptions for non-productive spaces such as crew quarters; 1 gross register ton is equal to a volume of 100 cubic feet. However, this calculation of GRT is quite complex. Gross register tonnage was replaced by *gross tonnage* in 1994 under the Tonnage Measurement convention of 1969, but is still a widely used term in the industry.

Net Register Tonnage (NRT)

NRT is the volume of cargo the vessel can carry; ie. the Gross Register Tonnage less the volume of spaces that will not hold cargo (e.g. engine compartment, helm station, crew spaces, etc). It represents the volume of the ship available for transporting freight or passengers. It was replaced by *net tonnage* in 1994, under the Tonnage Measurement convention of 1969.

Gross Tonnage (GT)

It refers to the volume of ship's all enclosed spaces (from keel to funnel) measured to the outside of the hull framing. It is always larger than *gross register tonnage*, though by how much depends on the vessel design. . It is a measurement of the enclosed spaces within a ship expressed in "tons"—a unit which was actually equivalent to 100 cubic feet.

As per IMO Convention, the correct term to use for tonnage measurement is GT, which is a function of the moulded volume of all enclosed spaces of the ship. It is calculated by using the formula GT = K.V., where V = total volume in m³ and K = a variable from 0.22 up to 0.32, depending on the ship's size (calculated by K = 0.2 + 0.02 × logV). GT is consequently a measure of the overall size of the ship.

Net Tonnage (NT)

It is based on a calculation of the volume of all cargo spaces of the ship. It indicates a vessel's earning space and is a function of the moulded volume of all cargo spaces of the ship.

10.4.5 Weight Measurements

Weight measurements are sometimes wrongly used for tonnage measurements. Nevertheless, weight measurement is an important concept for tankers and supertankers and hence the underlying principles like that of "Displacement" and "Archimedes Principle" need to be understood.

"**Displacement**" is the actual total weight of the vessel. It is often expressed in metric tons, and is calculated simply by multiplying the volume of the hull below the waterline (i.e. the volume of water it is displacing) by the density of the water. The density varies depending upon whether the vessel is in fresh or salt water, or is in the tropics, where water is warmer and hence less dense. The word "displacement" arises from the basic physical law "Archimedes Principle" which states that the weight of a floating object equates exactly to that of the water which would otherwise occupy the "hole in the water" displaced by the ship.

A Case Study— Field Development Scenario for a Marginal Field with Floating Production System and Sub-Sea Production System off Indian Offshore Shallow Water

In previous chapters, we have studied quite comprehensively the various attributes of the floating production systems and the sub-sea production systems. Now we will use all those knowledge and understanding to device an optimum field development scenario for a hypothetical marginal field at Indian Offshore waters or South East Asian waters in relatively shallow and calm water.

There are many possible ways to develop an offshore oil field. However, the best development plan is that which ensures its viability throughout the life of the field. Here, the scenario planning is important because we need to commit a production system in the early stage of field development when relatively little is known about the field and its reserves. This also warrants a good and comprehensive understanding of project management from comprehension to commissioning.

Any good development plan of a marginal field or early production system, based on the deployment of a floating production system, should possess the following characteristics:

- Minimum capital cost
- Early production feature in order to generate early cash flow
- Minimum technical risk by utilizing state-of-the-art components
- Major production components to be reusable

In addition, the field development concepts should be best suited to the environmental, physical, and socioeconomic concerns and due compliance must be assured.

FIELD DEVELOPMENT SCENARIOS: OPTIONS/ALTERNATIVES

The three probable field development scenarios, based on utilization of floating production systems for the field in focus at offshore India, are:

Option-A: FPSO Barge and Sub-sea Wells Development

Option-B: FPSO Tanker and Sub-sea Wells Development

Option-C: Semi-submersible FPS and Sub-sea Template Wells Development

While evaluating these alternatives, one must have sufficient, adequate and reliable knowledge of four things:

(a) Specific site and field conditions,

(b) Their production capability,

(c) System flexibility and

(d) Economical viability.

Under these four broader umbrellas, any field development scenario must have the adequate and detailed fact gathering, discussions and deliberations over the parameters as given below so as to evolve a comprehensive optimal field development plan:

- Field Parameters
- Field Development and Production System Component
  - A Floating Production System along with process facilities, utilities systems and facilities for firefighting and safety
  - Provisions to handle high pour point crude and waxing
  - Mooring systems and barge motions
  - Crude oil storage and offloading systems
  - Sub-sea production system comprising of sub-sea wells, risers, flow-lines, flexible lines and control umbilical.
- Production Profile
- Project Schedule
- Project Economics: Capital and Operating cost; in both the cases whether FPS company owned or FPS on lease.

Now on following pages, let's have detailed understanding and deliberations over these three options through above mentioned considerations, and as per the Facts and Figures mentioned in the Tables given at the back of the book. Figure mentioned in this chapter under various options and in Tables are just an approximations and hypothetical one given here to develop an understanding.

After these deliberations, we have found which FPS suits our requirement and why and this has been finally concluded and recommended.

Option-A: FIELD DEVELOPMENT SCENARIO THROUGH FPSO BARGE AND SUB-SEA WELLS DEVELOPMENT

The Floating Production, Storage and Offloading (FPSO) barge and sub-sea production system in this study is designed for a hypothetical, marginal oil field, offshore India or South East Asia, in relatively calm and shallow waters. The proposed barge-based FPSO system is aimed at fields likely to produce 3000 to 10,000 bopd. Suitable water depths range from 30 to 100 meters. Maximum 100 years wave height is not more than 12.5 meters.

There are certain definite advantages with this option. The barge-based production system is a very cost-effective method of developing marginal offshore fields in shallow waters and mild environments. It is suitable for extended well testing or early production of a larger field. These systems are also reusable. The barge-based FPSO has good capacity in terms of weight and volume for process equipment and storage. This gives a large degree of operational flexibility, as deck loads and storage are not constraints as they would be for semi-submersible rigs. The cost of a barge-based FPSO is lower than for a tanker-based FPSO, and the project duration to first oil production is also shorter.

However, barge-based FPSO system has some apparent disadvantages too. The large vessel motions limit it to shallow waters and mild/sheltered locations. The maximum number of wells it can handle is limited to eight, approximately. Drilling or work-over cannot be performed from the barge. The barge is fix-moored. In areas where the wind and waves are not predominantly coming from one direction, the barge may have to weathervane to reduce the motions and mooring loads. Other limitations of the barge-based FPSO system are very much related to the size of life of the field. As the field size and life increase, together with shallow water depth and mild environmental conditions, the fixed platform becomes a more viable alternative. The ability to process gas for export purposes is also limited. Thus, barge-based FPSO system is not suitable for marginal gas fields.

A1 FIELD PARAMETERS

At the very outset, it is important to know the field parameters. The major parameters assumed for the development scenario using the FPSO barge and sub-sea wells are:

Water Depth	:	60 m
Distance From Shore	:	100 km
Production Rate	:	10,000 bopd (maximum)
Production Per Well	:	2,500 bopd (maximum)
GOR	:	560 scf/bbl
Production Wells	:	4 (subsea wells)
Injection Wells	:	Nil
Pour Point	:	27°C
H_2S Content	:	230 ppm
CO_2 Content	:	3 percent
Field Life	:	Approximately 4 years

A2 PRODUCTION SYSTEM COMPONENTS

In order to exploit the field production, this option consists of the following major components:

- A FPSO Barge along with process facilities, utilities systems and facilities for firefighting and safety
- Provisions to handle high pour point crude and waxing
- Mooring systems and barge motions
- Crude oil storage and offloading systems
- Sub-sea production system comprising of sub-sea wells, risers, flow-lines, flexible lines and control umbilical.

Due to the marginal nature of the field, the total field development comprises only four sub-sea wells, of which two are located directly below the FPSO barge. The other two satellite wells are located approximately 2 kilometers from the FPSO barge. A shuttle barge will transport the crude to an onshore oil terminal at regular intervals. During normal production operations, the sub-sea wells will deliver the crude through the flexible flow lines and risers to the FPSO barge. The associated gas produced will first be treated and then will be used for fuel (for boiler and power generation, etc.). An excess amount of gas will be flared off via a remote vent stack. Produced water will be treated to reduce the oil content and other pollutants to an acceptable level before dumping overboard. The FPSO barge also has a 24-man accommodation and a heli-deck, located at the bow.

A2.1 The FPSO Barge

A2.1.1 *Barge Layout*

The proposed barge is a purpose-built vessel with a spoon bow and is fix-moored on location by eight mooring chains. It provides a considerable volume for oil storage, as well as being a platform for supporting the production, storage, export facilities and the living quarters. The barge is designed in accordance with American Bureau of Shipping's "Rules for Building and Classing Steel Barges for Offshore Service". It is classed as ABS A1 Oil Storage Barge.

The principal dimensions of the barge are as follows:

Length Overall	:	105 m
Width	:	21 m
Depth	:	9.25 m
Loaded Draft	:	6.5 m
Displacement	:	12,700 tones
Deadweight	:	10,000 DWT

(Figures close approximations)

The barge is provided with the Storage capacities as given below:

Cargo Oil	:	50,000 barrels
Water ballast	:	19,000 barrels
Produced Water	:	4,300 barrels
Potable Water	:	1,700 barrels

(Figures close approximations)

Let's understand the general arrangement of the elevation, top deck plan, and tank layout of the barge. A mono-pool is located at mid-ship to facilitate wireline operation. The helideck is located at the bow and provides a completely unobstructed landing area. The 24-man living quarters and the control room, made up of portable cabins, are located below the helideck. The process equipment and meter prover are mounted on a platform located near the mono-pool. This location is close to the center of gravity of the barge, where the vessel motions are the least severe. The platform is raised 3 meters above the top deck for ease of operation and maintenance, and to avoid waves splashing on equipment. The generators, boiler, and other utility equipment are also mounted on a platform at the stern. The platform is also raised 3 meters above the top deck, as required by codes and regulations.

The inlet manifold is located at the starboard side of the FPS barge. A 20-ton hydraulic crane is provided at port side for handling of supplies and lifting of equipment during maintenance. Below the deck, the largest part of the volume is devoted to oil storage. The tanks are subdivided into wing and central tanks. Segregated ballast is also incorporated, in accordance with the IMO recommendation for pollution prevention. The pump room is also located below deck, aft of all cargo oil tanks. It accommodates the offloading, stripping, seawater, ballast, and one of the fire pumps. The other fire pump is located on the deck to give a wide separation of these two emergency facilities.

A2.1.2 *Process Facilities*

The process is a three-stage, three-phase separation process. Separator specification can be understood from the separator specifications given at **Table 1** at the back of the book. Also refer **Table 3** for understanding complete list of equipment and facilities along with their dimensions and design conditions like capacities etc. During normal operation, crude from the sub-sea wells enters the first-stage (high-pressure) separator, where initial separation of oil, gas, and water is carried out. The high-pressure oil flows to the second-stage, low-pressure separator, and then to the atmospheric separator for further separation. Stabilized oil from the last-stage separator is pumped to storage. Gas from the HP separator is stripped of water and H_2S by the molecular sieve process before being used as fuel gas for power generation, heating, etc. Excess HP gas, together with LP gas from the second-and third-stage separators, flows to the high-and low-pressure knockout drums for removal of residual liquid prior to flaring. As the barge is not weathervaning, due to the high content of H_2S in the gas from the field under study, ground flare is not considered for safety reasons. The use of conventional flare booms is also not considered because this will interface with the offloading operation. Thus, the excess gas is sent to a remote stack, installed on a steel tripod, for flaring. Produced water from the separators flows to the plate coalescer for removal of suspended oil. Water discharged from the plate coalescer is pumped to the produced-water tanks for further settlement. Treated, produced water will contain no more than 35 ppm of oil before dumping overboard.

The process equipment proposed for the FPSO barge must operate satisfactorily on a constantly moving vessel. In order to minimize wave motion and primary turbulence, the separators are equipped with anti-surge baffles to reduce the effects of splashing and sloshing. The motion criteria specified for the processing equipment are 2-degree pitch and 12-degree roll, maximum. The FPSO barge and its mooring system are designed such that it does not exceed these motions by 5% of the time during operation. Further, to reduce the effects of motion, the separators and scrubbers, etc. are located longitudinally on the FPSO barge, as close to the center of gravity of the vessel as possible.

Because of the high pour point nature of the crude produced, the separation temperature is maintained at 42°C, which is above the pour point of the crude. All surface facilities handling high pour point crude are insulated to minimize heat losses. In addition, all process piping is heat-traced. This facilitates the restart of congealed crude in the piping after a prolonged shutdown.

It is desired that total amount of electric power required and its source of availability must be firmed before hand. The electrical power requirement was estimated to be 800 kW. 3 × 400 kW (i.e. 3 × 50%) diesel, dual-fuelled power generators are provided on board. To get an understanding, please refer to **Table 8** at the back of the book reflecting equipment wise power requirement for FPSO Barge based system.

A2.1.3 *Emergency Shutdown System (ESD)*

Alongside the process facilities, an emergency shutdown system is incorporated. It can be activated by any of the remote shutdown knobs located in the vicinity of the process equipment platform and at other strategic locations on the deck of the barge. The fire loop and H_2S detectors are part of the ESD system. Once the ESD system is activated, the following event occurs: (a) All production shutdown valves on the barge get closed, (b) The sub-sea down-hole safety valve gets closed (c) the firewater pumps and process area deluge gets started. In addition, equipment located in the pump room can be manually shut down from outside the room.

A2.1.4 *Firefighting and Safety System*

The fire safety system, complying with the attributes of SOLAS 1974, consists of the following:

- Fire detection system
- Gas detection system
- Heat/Smoke/Fire detection system
- Firefighting system
- Fire main header
- Fixed deck froth (foam) system
- Foam fire fighting system at heli-deck
- Two 20-man life rafts
- Life vests and other standard marine safety equipment

A2.1.5 *Other Utility Systems*

Besides the above systems, other utility systems are also provided. These are:

- Utility and instrument air supply
- Diesel fuel system
- Cooling water system
- Potable water system
- Ballast water system
- Bilge
- Chemical injection facilities
- Telecommunication
- Sewage treatment plant
- Well kill facilities
- Well-kick-off facilities

It is important to understand how the total lists of equipment on board FPSO Barge are being prepared and what is going to be the estimated operating weight of the process and utility facilities; because this estimated weight is important while choosing the dimensions and specifications of barge. In the case under consideration, total estimated operating weight of the process and utility facilities is 1257 tones. If we add the weight of living quarter, kill facilities and some other miscellaneous items, it works out to be 1950 tones. Please refer to **Table 3** at the back of the book to have a complete list of equipment (process + utilities + others) along with their representative dimensions, design consideration like capacities, pressure and temperature and prevailing indicative costs. This table will also help in understanding the underlying capital costs of the given project.

A2.2 High Pour Point Crude and Waxing

One of the major operational problems in producing high pour point crude occurs during the start-up of new facilities. The production facilities have to be heated to the process temperature (about 42°C) before allowing the crude to flow through the production system. One solution is to flow hot diesel oil through the process train before producing the crude. From an operational point of view, it is always a good practice to wash out residual high pour point crude with hot diesel oil from the process facilities before shutting them down. This practice should also be followed prior to shutdowns for routine maintenance or repair. This will facilitate quick start-up. To prevent wax buildup and congealing of crude in the flow lines, the following provisions are normally incorporated:

- The two satellite wells will have 3-inch annulus lines. Pigging operations can be carried out via these annulus lines and the production lines.
- The annulus and production lines are electrically heat-traced. This is a backup system. This operation will be activated only when there is an unplanned emergency shutdown and when the crude has congealed in the flow lines.
- Pour-point depressant will be injected at the wellhead to lower the pour point of the crude if the production system has to be shutdown.
- The shuttle barge will also be equipped with heated tanks to prevent congealing of the high pour point crude.

A2.3 Mooring System and Barge Motions

Environmental conditions in South East Asia are dominated by the NE and SW monsoon. It is mild compared to other parts of the world like that of Gulf of Mexico and North Sea. Hence, this sea environment is best suited for deployment of FPSO barge. There are two or three FPS barges currently operating in South East India Asia. The FPSO barge can be used for marginal field development or early production in western offshore India, especially in sheltered and relatively shallow waters where the winds and waves are prevailing from fixed directions for most of the year. Based on the operating experience of similar systems in South East Asia, and for economic reason, the barge is to be fix-moored on location. The mooring system is designed to keep the barge on station during the 100 years storm.

To understand what comprises the mooring system, for example, the mooring system planned for the FPSO barge in the given case consists of the following:

- Eight 73-millimeter diameters, 640-meter long ORQ chains, laid symmetrically at 22.5 degrees and 67.5 degrees to the barge's longitudinal axis. Each chain is pre-tensioned to 45 tones.
- Eight 30-tonne high-holding-power anchors
- Eight standard marker buoys
- Eight chain stoppers
- Four double-driven winches, with 50-tonne holding power for handling 44-millimeter wire rope.

Good software is available to analyze the barge motions in varied sea environment. Here in the given case, computer analysis shows that the barge motions are not excessive if it is deploying in South East Asian waters or offshore India with similar environmental conditions. Annual production downtime due to excessive pitch motion (greater than 2 degrees) is estimated to be 1.5%.

A2.4 Crude Oil Storage and Offloading System

The barge provides maximum crude oil storage of 50,000 barrels, which is equivalent to five days' production. Due to the high pour point nature of the crude, internal heating coils, using steam as heating medium, are installed at the bottom of each crude oil storage tank to maintain the temperature of crude at 42°C. Heating capacity is estimated to e 42 MM Btu/hr. To ensure a safe explosion level inside the storage tanks, an inert gas system is incorporated. The combustion type of inert gas generator (0.4 mmscf/d capacity) provides a continuous flow of inert gas to prevent air from being drawn into the storage space.

An 8000-DWT shuttle barge is planned to be used to transport the crude to shore. Offloading operations will be carried out with the shuttle barge moored along side the FPSO barge. Mooring of the shuttle barge can be achieved in sea conditions up to 1-meter significant wave height, and loading operations must be discontinued in seas greater than 2-meter significant wave height.

The FPSO barge is required to store production for a period of time which is equal to the round-trip time of the shuttle (4.5 days). This is calculated to be 45,000 barrels of oil. Considering maximum heel and fill percentage, the storage required on board the FPSO barge is 50,000 barrels. Excluding weather downtime, the shuttle barge will take about 3.5 days to make a round trip. Two diesel-driven loading pumps, 3000-barrel per hour capacity each, are provided on board the FPSO tanker. Recommended loading rate is 3000 barrels per hour. Loading time is about 17 hours. The crude is metered before being transferred to the shuttle barge.

Refer to **Table 10** at the back of the book to get a first hand idea of the various components (field and production data as well as operational parameters) required for finalizing the storage design.

A2.5 Sub-Sea Production System

A2.5.1 *Sub-Sea Wells*

There are four independent sub-sea satellite wells producing directly to the barge via sub-sea flow lines. Two wells are located directly beneath the barge, to enable wirelining to be carried out from the barge. The remaining two wells are located at a distance of 2 kilometers from the barge.

The wells are planned to be drilled from a jack-up drilling rig and utilize a mud line casing suspension system. The well completion will be for a sub-sea diver-assist wet tree assembly. The tree assembly is a diver-installed, wet satellite tree, 3-inch, 5000 psi WP single, with a 2-inch annulus side outlet on a tubing head. The tree is assembled from individual valves and components rather than utilizing a solid forged block in which the valves and bores are machined. Using individual valves reduces the tree costs and results in minimum delivery times.

The sub-sea production system also provides remote control for production, well control, annulus monitoring, flushing of all lines, and pigging capability. The selected control system for the sub-sea wells is a direct hydraulic system. There is no sub-sea monitoring or down-hole data acquisition system provided. Essentially, each valve which is required to be remotely operated is connected directly; via its control line, to the well control panel and hydraulic power unit located on the FPSO barge.

A sub-sea tree has many components. An outline specification for the proposed sub-sea tree is as follows (refer **Table 5**):

- Wet, diver-assist tree assembly
- Non-TFL—wireline compatible
- 2-1/16 inch side entry annulus
- 3-1/8 inch production bore
- Assembled form, individual API 5000 psi WP valves
- Lower master—manual
- Upper master—fail-safe hydraulic
- Swab valve—fail-safe hydraulic
- Production wing valves—fail-safe hydraulic
- Crossover valve—fail-safe hydraulic
- Manual annulus wing
- Manual flow line isolation valves

A2.5.2 *Risers and Flow Lines*

Sub-sea flow lines are the connecting link between the sub-sea trees and the FPSO barge. Risers and Flow Lines can be either flexible pipe or steel pipe. Normally, flexible pipes are selected over steel pipe for the flow line material, as flexible pipes has:

- Lower installed cost compared with double insulated rigid steel pipe
- Ease of installation and recovery, and can be reused
- Good insulation property, as the crude transported is of a high pour point nature

Riser configuration is important. So it is important to undertake a static analysis of various riser configurations using the appropriate software. Due to the relatively calm environment, a simple catenary riser configuration was statically analyzed. The catenary configuration gave acceptable riser shapes at the extreme motion envelope positions of the barge. It is planned to make the riser section integral with the flow lines and lay the flow lines from the tree directly to the barge. Diver-assist flange connections are utilized to connect the flow line to the trees. An emergency quick-disconnect system is provided on the barge, for the flow lines and control umbilical. This connection can be quickly disengaged in an emergency situation, i.e. breaking of mooring line(s), extreme weather, fire, etc.

A typical outline specification for the risers and flow lines (Production and Annulus lines and also the control Umbilical lines) can be understood from the components-specifications given at **Table 5** at the back of the book.

A3 PROJECT SCHEDULE

After undertaking all due deliberations and due-diligence over the major project activities, a project schedule is prepared which is separate and unique to a given project or development plan. In the present case under this development option, major project activities includes the engineering, construction, and offshore installation of the FPSO barge and its associated topside facilities, drilling and completion of four sub-sea wells, laying of sub-sea pipelines and riser, and offshore hookup and commissioning of the production system.

While developing the project schedule for this development, following assumptions were made:

- The FPSO barge was assumed to be purpose-built in one of the shipyards in India;
- The processing facilities, which are fabricated on skids or modules, are installed and tested on the FPSO barge before tow-out;
- The drilling and completion time required for each sub-sea well is about 40 days; and
- Finally, it was assumed that the shuttle barge will use an existing onshore oil terminal facility for offloading of the crude.

For this given development plan, as per the project schedule the first production from the field could begin about 15-1/2 months after project approval.

A4 PRODUCTION PROFILE

As mentioned above, production from the field will start 15-1/2 months after the start of the project. The production rate will peak very quickly at 10,000 bopd since all four sub-sea wells have been drilled before the arrival of the FPSO barge. The field is assumed to be producing at a maximum rate of 10,000 bopd for about 1-3/4 years, after which the production will decline to about 2500 bopd after 4 years of production, will decline to about 2500 bopd after 4 years of production. To allow for production downtime due to weather, the total number of days of crude production is assumed to be 330 days per year (about 10% production downtime). It is anticipated that at the end of the field's life, about 10 million barrels of oil will be produced.

A5 CAPITAL AND OPERATING COSTS

Now, after understanding all the concepts and attributes related to FPSO barge based FPS, let's now understand how the project economics gets worked out. I will suggest the students to go through the concept of "Capital Budgeting" and "Project Economics" to understand the feasibility report preparation exercise in totality. Working out the capital and operating costs is one very important exercise in this direction. While working out these two costs, two scenarios arise:

Case 1: Assume Oil Company owns FPSO Barge
Case II: Assume Oil Company gets FPSO Barge on Lease

To get an idea, how the capital and operating costs are worked out and what the various components of these costs are, refer to the example worked out under **Option-C** w.r.t. semi-submersible FPS option.

The estimated capital cost of this development is about US$ 47 million for Case I, and about US$ 30.25 million for Case II. The estimated annual operating costs at peak production for Case I and Case II are US$ 5.05 million and US$ 9.95 million respectively.

Also we need to workout quite clearly the exclusivities. For example, in the present case, the capital costs do not include or consider:

(a) Additional facilities or improvement work required at the onshore oil terminal to handle the oil produced from this marginal field.
(b) Additional support facilities or equipment such as tugboat, helicopter, and supply boat.
(c) Exploration and field abandonment costs were assumed to be included in the contingency cost.

One very important aspect to consider here is that at which place we are going to undertake the needed (if any) fabrication exercise: locally or at some fabrication yard elsewhere. The contract value against this is worked out accordingly. Here in this case, it is anticipated that fabrication of the FPSO barge and its associated topside facilities will be done locally by a shipyard for a total contract value of about US$ 14.02 million against these works inclusive of material cost.

One very crucial consideration here is where we are putting our abandonment cost. In the present case, the field abandonment cost is included in the contingency for capital cost.

The other important consideration is the assumptions made while working out the leasing rate and the components of the leasing equipment/systems. Further, it is important to know that in case of lease, whether the oil company is responsible for the operation and maintenance of the leased FPSO barge. In the present case, the oil company is responsible for the operation and maintenance of the leased FPSO barge, hence this component gets included in annual operating cost. The leasing rate of US\$ 12,250 per day (or US\$ 4.471 million per year), for this FPSO barge is calculated based on the following assumptions:

Leased period – 4 years

Lease company equity – 10%

Interest on debt – 14%

Lease company return on enquiry – 30%

Salvage value – 80% annually (i.e. 40% after 4 years)

I will advise students to work-out the capital and operating costs for this barge option on the similar lines as worked out under **Option-C** w.r.t. semi-submersible FPS option.

Option-B: **FIELD DEVELOPMENT SCENARIO THROUGH FPSO TANKER AND SUB-SEA WELLS DEVELOPMENT**

The FPSO tanker and sub-sea wells development consists of the FPSO tanker, six satellite sub-sea wells, and connecting flow lines/risers. The FPSO tanker is suitable for the development of marginal fields or for early production of a large field in medium water depths and in moderate environments.

The tanker-based production system is a flexible and cost-effective method of developing marginal offshore fields. It is also economical to use it for extended well testing or early production of a larger and more complex development. During recent years, there has been a steady increase in the acceptability of tanker-based floating production systems for marginal field development. This is due primarily to the need to provide less expensive solutions to the development, thereby making smaller fields more economically attractive to develop. In addition, there has been a steady improvement in the design and cost effectiveness of tanker mooring systems and sub-sea equipment.

B1 FIELD PARAMETERS

The production and field parameters assumed for the FPSO tanker and subsea wells development are:

Water Depth : 120 m
Distance From Shore : 160 km
Production Rate : 20,000 bopd (maximum)
Production Per Well : 2,500 bopd (maximum)
GOR : 560 scf/bbl
Production Wells : 6 (subsea wells)
Injection Wells : Nil
Pour Point : 27°C
H_2S Content : 230 ppm
CO_2 Content : 3 percent
Field Life : Approximately 5 years

B2 FIELD DEVELOPMENT COMPONENTS

Field development consists of the following major system components:

- A converted FPSO tanker
- Six satellite sub-sea wells
- Flexible flow lines, annulus lines, and control umbilical connecting the wells to the FPSO tanker
- Shuttle tankers

The overall system comprises a 130,000-DWT tanker, which is stern-moored to and weathervanes around a single point mooring system. The rigid yoke at the stern of the tanker is connected to the mooring column. The mooring column is anchored to the seabed by six chains. Flexible risers and control umbilical are enclosed in the J-tubes attached to the outside of the column. At the wave zone, they are totally enclosed in the mooring column. The process equipment is mounted at mid-ship, in order to reduce the influence of the vessel motions.

The six satellite sub-sea wells produce the oil to the FPSO tanker through the flexible flow lines via the mooring column. The flow lines are manifolded above water before flowing through the multi-pass swivel. Most of the tanks below the deck are used for crude oil storage and water ballast systems. The FPSO tanker has a maximum 700,000 barrels of oil storage capacity. Oil, gas, and produced water are separated in the separators. The crude oil is temporarily stored in the heated storage tanks. The crude oil will be metered before pumping to a shuttle tanker. The gas will be used as fuel or will be flared-off via the ground flare. Produced water will first be treated before dumping overboard. The FPSO tanker also has accommodation facilities for 40 men, and a helideck.

The following paragraphs describe the recommended system components for this field development.

Description of major components and operations:

B2.1 FPSO Tanker

B2.1.1 *The Converted Tanker*

The FPSO tanker is a converted 130,000-DWT tanker. A typical 130,000-DWT tanker will have the following principal dimensions:

Length Overall	: 273.0 m
Length between Perpendiculars	: 260.0 m
Molded Breadth	: 42.0 m
Molded Depth	: 23.50 m
Loaded Draft	: 16.58 m
Light Draft	: 2.97 m
Displacement (Loaded)	: 154,000 tones
Light weight	: 24,000 tones
Deadweight	: 130,000 tones

(Figures close approximations)

An existing tanker is chosen rather than a purpose-built one. This is due to the fact that, with the depressed market for over 100,000-DWT tankers, acquiring and converting an existing 130,000-DWT tanker is less expensive than a smaller, purpose-built tanker. Furthermore, a large tanker has better motion characteristics than a smaller tanker. This will lead to lower weather downtime. Further, the tanker chosen for conversion should preferably be less than 10 years old. This is because older tankers will not likely have segregated ballast, crude oil washing, and inert gas systems, as required by regulatory authorities and classification societies. To retrofit these systems will be costly. Also, equipment on an older tanker may be obsolete and spare parts for maintenance purposes may not be available.

The tanker is permanently moored at the stern by a turret column. The tanker may stay on station for as long as 12 years, provided annual and special surveys are carried out. This requires the tanker to be structurally sound. Impressed current system is installed to protect the underwater hull. In addition, crude oil, water ballast, and produced water tanks are protected by the sacrificial anodes.

The converted tanker is to be de-engined, deregistered, and reclassified as a fixed structure. This is the trend for most of the FPSO's installed since 1970's. As such, a full marine crew is not required. A 40-man living quarters is sufficient. Here in this case, no expansion of existing living quarters/deck house is anticipated. A doctor room and a recreation room are incorporated in the living quarters.

Let's understand the general arrangement of equipment on the converted tanker.

A helideck is located just forward of the deck house. The ideal location of the helideck will be at the stern. However, the rigid arm extension attaching the tanker to the mooring column occupies most of the space at the stern. It is not practical to raise the helideck above the rigid arm. The process equipment is mounted on a platform located at midship. This location is close to the center of gravity of the tanker, where the tanker motions are least severe. The platform is raised 3 meters above deck for ease of operation and maintenance. Meter facilities are also mounted on this platform. A ground flare is located at the forward section for disposal of excess gas. Two 20 tones, 15-meter boom, pedestal-mounted cranes are located between the process platform and the ground flare for handling of materials from the supply boats and also for lifting the equipment during maintenance. A gantry is provided on the forecastle deck for handling of loading hose since tandem-to-tandem offloading is contemplated. The living quarters/deck house is located at the stern. Below deck, the largest part of the volume is devoted to crude oil storage. The tanks are subdivided into wing and central tanks. Segregated ballast is also incorporated, in accordance with the IMO recommendation for pollution prevention.

B2.2 Process Facilities

The process is three-stage, three-phase separation process. Just to develop an understanding, separator specification can be referred at **Table 1** at the back of the book. Also refer **Table 2** for understanding complete list of equipment and facilities along with their dimensions and design conditions like capacities etc.

During normal operation, fluid from each sub-sea well is directed up through the risers, into the manifold located on the mooring column, and commingled into the 8-inch production header. Fluid from the header is taken off by two 6-inch lines, passed through two levels of the swivel and delivered to the first-stage (high-pressure) separator. Initial separation of oil, gas, and water is carried out. The high-pressure oil flows to the second stage-(low-pressure) separator, and then to the atmospheric separator for further separation. Stabilized oil from the last-stage separator is pumped to storage. Gas from the HP separator is stripped of water and H_2S in the molecular sieves, prior to being used as fuel gas for power generation and boilers. Excess HP gas, together with LP gas from the second-and third-stage separators, flows to the high-and low-pressure knockout drums for removal of residual liquid prior to flaring. As the FPSO tanker is able to weathervane and is moored at the stern, a ground flare is used for flaring. It is located at the forward section so that it is always downwind of the living quarters/deck house. In this way, this arrangement reduces the H2S hazard. Produced water from the separators flows to the plate coalescer for removal of suspended oil. Water discharged from the plate coalescer is pumped to the produced water tanks for further settlement. Produced water, after treatment, will contain no more than 15 ppm of oil before it is disposed to the sea.

Individual wells can be selectively tested by diverting the flow to the 6-inch test header, through the swivel, and delivered to the test separator. Separation of oil and water is carried out and their flow rates are measured. After testing, oil is fed to the low-pressure atmospheric separators for further separation. Produced water and gas are sent to the plate coalescer and knockout drums, respectively, for further processing.

The process equipment is designed to operate satisfactorily on a constantly moving vessel. In order to minimize wave motions and turbulence, separators are equipped with anti-surge baffles to reduce the effects of splashing and sloshing. They are designed to function effectively to a maximum of 2-degree pitch and 12-degree roll tanker motions.

Due to the high pour point nature of the produced crude, the separation temperature is maintained at 42°C, which is above the pour point of the crude. All surface facilities handling high pour point crude are insulated to minimize heat losses. In addition, all process piping is heat-traced. This will facilitate the restart of congealed crude in the piping after a prolonged shutdown.

It is desired that total amount of electric power required and its source of availability must be planned before hand. The electrical power requirement is estimated to be 1000 kW. The selected tanker's existing generators can deliver 150% of the above requirement. To get an understanding, please refer to **Table 9** at the back of the book reflecting equipment wise power requirement for FPSO Tanker based system.

B2.3 Emergency Shutdown System (ESD)

Further, it is prudent that alongside the process facilities, an Emergency Shutdown (ESD) system is incorporated. ESD can be activated by any of the remote shutdown knobs located in the vicinity of the process equipment platform and at other strategic locations on the deck of the tanker. The fire loop and H_2S detectors are part of the ESD system. Once the ESD system is activated, the following events will occur: (a) All production shutdown valves on the tanker get closed (b) the sub-sea down-hole safety valve gets closed (c) The firewater pumps and process area deluge gets started. In addition, equipment located in the machinery room can be manually shutdown from outside the room.

B2.4 Firefighting and Safety System

The fire safety system, complying with the attributes of SOLAS 1974, consists of the following:

- Fire detection system
- Gas detection system
- Firefighting system
- Fire main header
- Fixed deck froth (foam) system
- Two 30-man life rafts
- Life vests and other standard marine safety equipment

B2.5 Other Utility Systems

Besides the above systems, other utility systems are also provided. These are:

- Utility and instrument air supply
- Diesel fuel system
- Cooling water system
- Potable water system
- Ballast water system
- Bilge
- Chemical injection facilities
- Telecommunication equipment
- Sewage treatment plan
- Well kill facilities

It is important to understand how the total lists of equipment on board FPSO Tanker are being prepared and what is going to be the estimated operating weight of the process and utility facilities; because these estimated weights are important while choosing the dimensions and specifications of tanker. In the case under consideration, total estimated operating weight of the process and utility facilities is 1380 tonnes. If we add the weight of living quarter, kill facilities and some other miscellaneous items, it works out to be 1850 tonnes. Please refer to **Table 2** at the back of the book to have a complete list of equipment (process + utilities + others) along with their representative dimensions and prevailing indicative costs. This table will also help in understanding the underlying capital costs of the given project.

B2.6 Crude Oil Storage and Offloading System

The FPSO tanker is required to store production for a period of time which is equal to the round-trip time of the shuttle tanker (5.5 days). This is calculated to be 110,000 barrels of oil. Considering maximum heel and fill percentages, the storage required on board the FPSO tanker is 122,000 barrels. A 130,000—DWT tanker is proposed to be used as the FPSO tanker for economic reasons and better motion characteristics.

The FPSO tanker has maximum crude oil storage of 700,000 barrels. To ensure a safe explosion level inside the storage tanks, a continuous flow of inert gas is required to prevent air from being drawn into the storage space. The tanker's existing inert gas generator, together with the exhaust from the boiler, is assumed to be sufficient for this purpose.

Storage capacities for this proposed tanker are as follows:

Crude Oil	:	700,000 bbls
Water Ballast	:	235,000 bbls
Slop tank	:	23,000 bbls
Produced Water	:	4,000 bbls
Miscellaneous	:	60,000 bbls

(Figures close approximations)

Refer to **Table 10** at the back of the book to get a first hand idea of the various components (field and production data as well as operational parameters) required for finalizing the storage design.

Due to the high pour point nature of the crude, internal heating coils, using steam as the heating medium, are installed at the bottom of each crude oil storage tank to maintain the temperature of crude at 42°C. Heating capacity is estimated to be 100 MMBtu/hr which works out to be adequate against the given requirement.

One shuttle tanker is proposed. It shall have a minimum storage capacity of 110,000 barrels of oil, or approximately 17,000 DWT. However, for economic reasons, a 50,000-DWT tanker is proposed, as the unit transportation cost will be the lowest. Since the shuttle is over-sized, it should not be dedicated to serve the field in focus alone. It should also serve other fields in the vicinity. A 50,000-DWT shuttle tanker can transport approximately 375,000 barrels of oil. Based on a loading time of one day, the recommended loading rate is: $375,000/24 = 15,625$ bbl/hr, say 16,000 bbl/hr. Tandem offloading operation will be carried out with the shuttle tanker moored bow to the FPSO tanker. Tandem rather than side-by-side offloading is proposed here, because the offloading operation can be carried out in higher sea states. It is envisaged that mooring of the shuttle tanker can be carried out in seas up to 4-meter significant wave height, and loading operations have to be discontinued in seas greater than 5-meter significant wave height in western offshore India.

Excluding weather downtime, the shuttle tanker will take about 3.5 days to make round trip to say vadinar port (Gujarat, India). The proposed loading rate is 16,000 barrels per hour. Against this requirement, the tanker's existing loading pumps are assumed to have sufficient capacity. However, if requirement is more, then we should also consider the provisions of additional loading pumps. The crude is pumped to the meter prover on board prior to discharge to the shuttle tanker.

B2.7 Mooring System and Tanker Motion

The environmental conditions prevailing offshore India are not mild. A single-point mooring system is selected so that the tanker can weathervane and take up positions of least resistance to wind, wave, and current.

In the given field and sea-environment scenario, SBM's turret mooring column system having quick-disconnect feature has sounded the best and hence selected. It is similar to BHP's Jabiru mooring system. One of the advantages of this mooring system, compared with other conventional mooring column designs, is that it can be adapted to different water depths (± 20 meters) by changing the mooring chain length or size.

The tanker is stern moored to the turret mooring column via the rigid arm extension. The riser column is anchored to the seabed by six stud-link chains. It is 100 meters long and mostly 6 meters in diameter; thins to about 3 meters in the splash zone to minimize wave loads. It is also compartmentalized. The two lower compartments are allocated for solid and water ballast. Flexible risers and hydraulic umbilical are enclosed in J-tubes which are attached to the outside of the riser column. At the wave zone, they are totally enclosed in the column. The flow lines are manifolded above water before passing through the multipass swivel.

Good software is available to analyze the barge motions in varied sea environment. Here in the given case, computer analysis shows that the barge motions are not excessive if it is deploying in Indian waters, offshore Bombay. Annual production downtime due to excessive pitch motion (greater than 2 degrees) is estimated to be less than 1%.

B2.8 Sub-Sea Production System

B2.8.1 *Sub-Sea Wells and Sub-Sea Tree*

There are generally four to six sub-sea satellite wells producing directly, via sub-sea flow lines, to the FPSO tanker. As the selected tanker mooring system could support the required number of production risers, a sub-sea manifold to be used in conjunction with the satellite wells is not proposed here. Further, owing to the system's weathervaning capability, clustered well arrangements to enable minor well work-over, with access from the FPSO tanker, are also not proposed.

The completion system is a conventional, guideline, sub-sea wellhead system. It enables remote installation and testing of the wellhead and completion string without the aid of divers.

The tree assembly is a wet tree assembly designed to be installed and tested with minimum diver assistance. Proposed diver tasks are associated with the connection of the sub-sea flow lines and control umbilical to the tree. The sub-sea tree system provides remote control for production, well control, annulus monitoring, flushing of all lines, and pigging capability.

The selected sub-sea control system is direct hydraulic. Each sub-sea valve actuator is connected, via a control line, to the hydraulic control panel located below the swivel on the mooring system. An independent data acquisition system provides wellhead and bottom-hole temperature and pressure indications. The sub-sea control equipment is installed inside the mooring column. The Hydraulic Power Unit (HPU) is installed before the swivel. This results in the need for only very low-pressure hydraulic line, and electrical power and signal lines from the FPSO tanker.

The multi-pass swivel must accommodate redundant 3000 psi and redundant 5000 psi hydraulic lines, in addition to the redundant electrical power and signal lines. Electrical power is single phase, 240 VAC, while the signal lines carry Frequency Shift Keyed (FSK) digital two-way communication from the control room panels to the multiplex controllers attached to the riser.

Multiplex electro hydraulic controllers are attached to the risers below the swivel but above sea level. This type of controller is necessary to minimize the complexity of the electrical swivel which would otherwise require a large number of electrical lines.

A sub-sea tree has many components. An outline specification for the proposed sub-sea tree is as follows (refer **Table 6**):

- Wet, diver-assist tree assembly
- Non-TFL – wireline compatible
- Hydraulic wellhead connector – 16-3/4 inch × 5000 psi WP
- 3-1/8" × 2-1/16" dual-bore, monoblock master valve block with integral master and swab valves
- Lower manual production master
- Upper production master – hydraulic fail-safe close
- Single annulus master – hydraulic fail-safe close
- Swab valves – hydraulic fail-safe close
- Crossover valve — fail-safe open
- Tie-back mandrel for vertical access to annulus and production bores
- Tree cap
- Manual flow line isolation valves

B2.8.2 *Risers and Flow Lines*

Sub-sea flow lines connect the mooring column and sub-sea trees. Each tree has a single annulus and production flow line (76-millimeter, 3 inch ID), as well as a control umbilical. All lines are of the flexible hose type rather than steel pipes, primarily for ease of installation and their reusable feature.

The turret mooring system necessitates the requirement of a "Lazy S" riser configuration. This configuration enables the risers to be integral with the flow line segment. The riser is connected to the mooring system first, and then the flow line is laid to the tree for a second-end connection.

A typical outline specification for the risers and flow lines (Production and Annulus lines and also the control Umbilical lines) can be understood from the component's specifications given at **Table 6** at the back of the book.

B2.9 High Pour Point and Waxing

One of the major operational problems in producing high pour point crude occurs during the start-up of new facilities. The production facilities have to be heated to the process temperature (about 42°C) before the crude is allowed to flow through the production system. This can be done by pumping hot diesel oil through the process train.

It is proposed to displace residual high pour point crude with diesel oil from the process facilities before shutting them down. This practice should also be followed prior to shutdowns for routine maintenance or repair. This will facilitate quick start-up.

To prevent wax buildup and congealing of crude in the flow lines, the following provisions are normally incorporated:

- The wells will have 3-inch annulus lines. Round-trip pigging operations can be carried out via these annulus and the production lines.
- The annulus and production lines are electrically heat-traced. This is a backup system. This operation will be activated only when there is an unplanned emergency shutdown and when the crude has congealed in the flow lines.
- Pour-point depressant will be injected at the wellhead to lower the pour point of the crude if the production system has to be shut down.
- To prevent wax buildup and congealing of crude in the storage tanks, all tanks (both FPSO and shuttle tanker) are equipped with steam-heating coils to keep the crude storage temperature at 42°C. In addition, the crude oil washing system allows hot oil to be sprayed on the storage tanks to remove any buildup of congealed crude.

It is anticipated that wells will be requiring frequent wirelining and well maintenance. These operations are supposed to be carried out from a semi-submersible rig or drill ship.

B3 PROJECT SCHEDULE

After undertaking all due deliberations and due diligence over the major project activities as given below, a project schedule is prepared which is separate and unique to a given project or development plan. In the present case under this development option, major project activities are:

(a) Preparation of tender documents and specification for an FPSO tanker, bid evaluation, and contract award;
(b) Selection of suitable tanker for conversion;
(c) Engineering, Design, material procurement, supply and fabrication of topside facilities for the FPSO tanker, overall system integration, hooking-up, pre-commissioning and commissioning;
(d) Engineering, Design, fabrication, and offshore installation of the Single-point Mooring System (SPM);
(e) Drilling and well completion of sub-sea wells;
(f) Laying of pipelines and control umbilical from the wells to the FPSO tanker; and
(g) Offshore hookup of total production facilities and commissioning of the floating production system.

While developing the project schedule for this development, following assumptions were made:

(a) Drilling and completion of each well take approximately 40 days;
(b) Tanker conversion and outfitting are carried out in a nearby Indian shipyard;
(c) Design and fabrication of SPM require approximately 13 months; and
(d) All wells are completed before tow-out of the FPSO tanker at location

Based on the above project activities and assumptions, a project schedule has been drawn indicating that first oil production from the field can begin 20 months after project approval.

B4 PRODUCTION PROFILE

In this development, it was assumed that all six wells have been drilled and completed before the arrival of the FPSO tanker. The production rate will peak to 15,000 bopd in a relatively short time period after the start of production. This plateau rate was assumed to continue for about 1-3/4 years, followed by a gradual decline in production until a production rate of about 3000 bopd is reached 5 years after the production starts. Beyond this point, it was assumed that the production would stop as revenue from this production rate will not be sufficient to cover the operating and maintenance costs of the floating production system.

To allow for production downtime, it is assumed the field is producing about 330 days per year. An estimated 18 million barrels will be produced at the end of the life of the field.

B5 CAPITAL AND OPERATING COSTS

Now, after understanding all the concepts and attributes related to FPSO tanker based FPS, let's now understand how the project economics gets worked out. I will suggest the students to go through the concept of "Capital Budgeting" and "Project Economics" to understand the feasibility report preparation exercise in totality. Working out the capital and operating costs is one very important exercise in this direction. While working out these two costs, two scenarios arise:

Case 1: Assume Oil Company owns FPSO Tanker

Case II: Assume Oil Company gets FPSO Tanker on Lease for a fix term

To get an idea, how the capital and operating costs are worked out and what the various components of these costs are, refer to the example worked out under **Option-C** w.r.t. semi-submersible FPS option.

The estimated capital cost of this development is about US$ 105.50 million for Case I, and about US$ 70.25 million for Case II. The estimated annual operating costs at peak production for Case I and Case II are US$ 9.45 million for Case I and US$ 18.75 million for Case II respectively.

One very important aspect to consider here is that at which place we are going to undertake the needed (if any) fabrication exercise: locally or at some fabrication yard elsewhere. The contract value against this is worked out accordingly. Here in this case, it is anticipated that fabrication of the FPSO tanker and fabrication of the SPM system, process and utility facilities will be done locally by a shipyard for a total contract value of about US$ 21.39 million against these works inclusive of material cost.

Also we need to workout quite clearly the exclusivities. Like for example, in the present case, the capital costs do not include or consider:

(a) Additional facilities or improvement work required at the onshore oil terminal to handle the oil produced from this marginal field
(b) Additional support facilities or equipment
(c) Exploration cost

It was assumed that the field abandonment cost is included in the contingency for capital cost. It was also assumed that the oil company (or operator of the field) is responsible for the operation and maintenance of the leased FPSO tanker.

One very crucial consideration here is where we are putting our abandonment cost. In the present case, the field abandonment cost is included in the contingency for capital cost.

The other important consideration is the assumptions made while working out the leasing rate and the components of the leasing equipment/systems. Further, it is important to know that in case of lease, whether the oil company is responsible for the operation and maintenance of the leased FPSO barge. In the present case, the oil company is responsible for the operation and maintenance of the leased FPSO barge, hence this component gets included in annual operating cost. The leasing rate of $ 23,200 per day (or US$ 8.47 million per year) for this FPSO tanker is calculated based on the following assumptions:

Leased period – 5 years
Lease company equity – 10%
Interest on debt – 14%
Lease company return on enquiry – 30%
Salvage value – 80% annually (i.e. 33% after 5 years)

I will advise students to work-out the capital and operating costs for this barge option on the similar lines as worked out under **Option-C** w.r.t. semi-submersible FPS option.

B6 COMPARATIVE ADVANTAGE: TANKER-BASED SYSTEM

For most applications, the tanker-based system offers a number of advantages over the semi-submersible-based system. In particular, the ability to store large volumes of oil and not be limited by deck load capacity gives the tanker system a large degree of operational flexibility than the semi-submersible system. Also, in general, tankers are more readily available at lower cost in today's depressed tanker market. It is only in the most hostile environments where vessel motions become the limiting factor; the better motion responses of the semi-submersible make it a viable option. Also, unlike the semi-submersible, the tanker-based FPS's normally do not have continuous well-access capability.

The proposed FPSO tanker development considers the use of satellite sub-sea wells. With the use of satellite trees, it is possible to use successful exploration wells as producers, provided a full casing program is completed at the exploration stage. This provides benefits, both in terms of development cost and shorter development time.

The potential ability to relocate this production facility to another location on field depletion is an important consideration. Depending on the similarity of the characteristics of the two locations, it is conceivably possible to reuse the FPSO tanker and its mooring system with minimal modification. The down-hole equipment would probably be abandoned, although the sub-sea trees, control systems, and flexible flow lines may be reused after major refurbishment.

B7 AREAS OF CONCERN: TANKER-BASED SYSTEM

Tanker-based systems do have some areas of concern that need to be acknowledged before we decide upon this option. Some of the areas of concern in the proposed tanker-based system are:

- Design of high-pressure, multi-pass swivel is rather difficult and complex
- The requirement of a reusable riser and mooring system for the FPSO tanker in varying water depths is not field proven.
- A above-water manifold is desired for the proposed mooring system, based on production from six wells. Should more wells than six be considered, or if water injection/gas lift are required, consideration should be given to manifold the flow lines sub-sea.
- The current FPSO tanker system is aimed at fields likely to produce 5000 to 40,000 bopd. At present, a single-point mooring (SPM) and riser system, in 180-meter water depth, is at the limit of state of the art. An

FPSO tanker in water depths greater than 150 meters will require another type of mooring and riser system.

- Other limitations of the tanker-based system are very much related to the size and life of the field. As the field size and life increases, platforms become a more viable alternative because of the lower risk of production downtime, the ability to handle more wells, and if an export pipeline is used, it will no longer rely on the weather-sensitive, offshore loading operations.

Option-C: FIELD DEVELOPMENT SCENARIO THROUGH SEMI-SUBMERSIBLE FPS AND TEMPLATE WELLS DEVELOPMENT

The semi-submersible Floating Production System (FPS) is ideally suited for development of marginal field or early production of fields located in medium to very deep waters and in moderate to harsh environments. A semi-submersible FPS is well suited for several field applications over its life expectancy, because of its ability to provide good stability over a large range of environmental conditions. Also, its mooring system is capable of handling a large range of water depths with minimum or no system modifications.

The semi-submersible-based floating production system has several potential applications for use offshore India like for Early Production Development, for Marginal Field Developments and for Extended Well Tests.

C1 FIELD PARAMETERS

The production and field parameters assumed for this field development through Semi-Submersible FPS and Template Wells development are as follows:

Water Depth	: 120 m
Distance From Shore	: 160 km
Production Rate	: 20,000 bopd (maximum)
Production Per Well	: 2,500 bopd (maximum)
GOR	: 560 scf/bbl
Production Wells	: 6 (template wells)
Injection Wells	: Nil
Pour Point	: 27°C
H_2S Content	: 230 ppm
CO_2 Content	: 3 percent
Field Life	: Approximately 5 years

C2 FIELD DEVELOPMENT COMPONENTS

Field development consists of the following major system components:

- A converted semi-submersible vessel with process facilities on board
- A sub-sea well template with an integral sub-sea manifold
- Flexible flow lines/risers connecting the sub-sea manifold to the FPS
- A CALM loading buoy and temporary storage/shuttle tankers (Offloading facilities, Shuttle Tankers and Storage)

The proposed sub-sea production system comprises an eight-slot, six well templates with an integral manifold. The template is located directly below the semi-submersible FPS. The sub-sea Christmas trees and manifold can be operated by a direct hydraulic control system. The group production line, test line, and annulus line from the manifold are connected to the riser bases. These lines and the control umbilical then assume a "Steep S" configuration, terminating off the semi-submersible. The flexible risers have Quick Disconnect (QC-DC) connectors located on the side of the main deck. Down-hole servicing will be carried out from the semi-submersible rig, using a rigid work-over riser.

The crude from the wells flows to the topside facilities via the manifold and the flexible production riser. The topside facilities (process facilities and associated utilities) are designed to support the production of about 20,000 bopd of crude oil. The crude oil which is separated from the separators will be pumped directly to a shuttle tanker, moored about 1-1/2 kilometers away to a CALM loading buoy. The proposed semi-submersible FPS is also equipped with at least 40-man living quarters and a helideck.

Description of major system components and operations:

C2.1 Semi-Submersible FPS

The most probable type of semi-submersible FPS is a converted, medium-size (2000 to 3000 tones variable deck load) semi-submersible drilling rig. A converted semi-submersible unit is preferred over a purpose-built unit as purpose-built unit is quite expensive. Further, the choice depends upon what size of semi-submersible drilling rigs which are suitable for conversion (small-, mid- or large-size) are currently available in the world-market.

The required variable deck load in tones for the semi-submersible FPS is normally calculated with the following components like Added Operating Deck Weight (process, utilities, production risers; say 868 tones), Onboard Fixed Weight for Well Work-over Equipment (BOP's, risers; say 314 tones),

Additional Operating Loads for Major Well Work-over (working loads and added consumables; say 584 tones), thereby calculating the total maximum variable deck load (say 1766 tones) by adding all those component weights.

The converted semi-submersible drilling rig is chosen here provides the following:

- A safe, stable platform with moorage and motion characteristics adequate to survive the 100 years storm and to remain fully operational in a 10 years storm.
- Capable of supporting an operating deck load (process, storage, and work-over equipment) of 1766 tones.
- Enables simultaneous production, processing, and offloading while performing minor (wirelining) well maintenance operations.
- Enables major well work-over activities with adequate load-carrying capacities for process and utility systems and anticipated well work-over consumables.

C2.2 Semi-Submersible Rig (to be converted to semi-submersible FPS)

The chosen rig is a self-propelled, twin-hull semi-submersible with a total of eight cylindrical stability columns and a transverse tubular truss arrangement. Each pontoon is subdivided by bulkheads to provide 15 tanks for fuel oil, drill water, and ballast water. The corner columns contain two chain locks and bulk storage tanks, with the lower columns, one on each pontoon, be modified to provide a crude oil storage capacity of about 5000 barrels.

The rig's deck supports the deck house, drilling equipment, mud pits, pipe racks, diesel generators, mooring winches, miscellaneous stores, and machinery equipment. An upper deck, above the accommodation unit and mud tanks, supports additional accommodation units if desired, the pilot house and radio room, shale shaker house and helideck. The main deck area is 69.2 × 61 meters (227 × 200 feet) of which 350 square meters (3900 square feet) are required for process, metering, and offloading systems for FPS service in this study. It is proposed that the required area be made available by decreasing the pipe rack area to half its normal size required for drilling operations. This approach involves the last amount of rig conversion work.

It is desired that the selected rig be in good structural condition, be less than 10 yeas old, and have full and appropriate classification certification in force at time of selection, (Det norske Veritas (DnV), or ABS or other equivalent classification authority in India), so as to be used as semi-submersible drilling platform for unlimited applications at the given site (say off the coast of India).

It is obvious that after conversion to an FPS, the classification be amended for the appropriate service and the rig should comply with the local government's appropriate codes, requirements, and regulations.

System integrity needs to be established post-conversion for the given applicability. In general, the rig, as supplied for FPS service, has to undergo system-by-system check, as given below for example, for being in good condition and wherever it is not, suitable modification exercise as per the need has to be undertaken.

- *Ballast System:* Good condition, No modifications required
- *Mooring System:* No modifications anticipated if selected rig has eight 76-milimeter (3-inch) diameter ORQ chains.
- *Propulsion System:* No modifications required
- *Accommodation Unit:* No modifications required
- *Accommodation Support Systems:* No modifications required
- *Drilling Equipment:* No modifications required. System used for major and minor well work-over.
- *Power Generation:* Converting existing diesel drivers to dual fuel, gas and diesel. No additional power utilities required for process and offloading systems.
- *Drilling Support Equipment:* Well test and well logging services are normally leased for drilling operations and are not required for FPS service.
- *Mud System:* No modifications required. Some of the mud and cement pumps and their deck tanks could be removed, if required, to accommodate more deck load.
- *Drill String and Tools:* Drill collars and drill pipe removed for weight savings.
- *Sub-sea Equipment:* BOP stack, handling, control, and test systems remain on board. Marine riser stored on shore.
- *Communications and Navigation:* Systems suitable for desired communication systems working in the area.
- *Safety Equipment:* No modification required. Additional firefighting systems added for process unit.

C2.3 Major Conversion Tasks

After undertaking the above explained exercise, major conversion tasks are identified incorporating all those additional things or systems which are not there on the original rig or deleting some equipment/system that are not required or modifying existing systems/equipment as per the given field development need. Here in the instant case, major rig conversion tasks include providing a crude oil storage system and associated pumping and venting systems, installation and hookup of process and utility systems, removal of two well test flare booms and installation of two larger production flare booms, fabrication of the riser support frame and installation of female half of riser connectors and associated pipe work, and installation of the sub-sea template control station. The total weight of the required process and utilities is estimated at 860 tones, which is accommodated on the deck by a corresponding weight removal of drill pipe, sack material, and other redundant drilling equipment.

Further, it is also important to plan for the lay-out of the equipment that has to be added on the converted rig. For example, here in the instant case, the process, metering facilities, custody transfer pumps are to be positioned on the aft main deck, at a maximum distance from the accommodations and helideck (portion of pipe rack area). Equipment modules are to be welded directly to the main deck. The flare booms are positioned on the two aft columns at main deck level. The riser terminations are located at the aft end, near the centerline and in close proximity to the process equipment.

C2.4 Process, Utility and Storage Systems

Process, Utility and Storage Systems for any FPS is designed on the basis of very basic production and field data such as how much oil and gas FPS is supposed to handle (say 20,000 bopd and 15 MMscfd gas), what is the API gravity of crude oil (say 38° API), how much is the CO_2 and H_2S ppm (say 3% CO_2 and 230 ppm H_2S), what is pour point of crude (say 27°C), how much storage capacity is required (say minimum 5000-bopd storage capacity) and like these other production, reservoir, well and sea parameters.

The system is designed to process and stabilize the crude on the semi-submersible and then pump the stabilized crude directly to a shuttle/storage tanker with some minimum storage capacity provided on the semi-submersible FPS to accommodate storage tanker change-outs. Produced water is treated on board the FPS before disposing to sea. The majority of the produced gas is flared on board the semi-submersible via flare booms. Fuel gas for the dual-fuel (gas and diesel) power generation dives and for the inert gas generation system is treated for H_2S removal.

To develop an understanding over how Process, Utility and Storage Systems are planned, designed and specified (represented), refer **Table 4** at the back of the book wherein a complete list of equipment planned for the given field development through semi-submersible FPS is listed out along-with the dimensions, specifications, design conditions and approximate costs involved.

C2.5 Mooring System and Vessel Motions

Careful inspection of the condition of the sea as well as of the mooring system is required whenever we go for selecting a particular rig. It is important to have a proper and due cognizance of prevailing sea conditions of the chosen location such as Maximum Wave Height, Associated Wave Period, Wind Speed (1-minute Storm) and Surface Current in order to ensure a better survival conditions for a mooring system and then compare the same with the 100 years storm conditions to establish its suitability and viability.

For example, the chosen semi-submersible FPS will be fix-moored on location by eight 76-millimeter (3-inch) diameter ORQ chains, for a sea condition as given below:

Maximum Wave Height : 30.5 m (100 ft)
Associated Wave Period : 14 seconds
Wind Speed (1-minute Storm) : 100 knots
Surface Current : 3 knots

Semi-submersible rigs have better motion characteristics than tankers. Specification of rig and vessel show that the vessel motions are not excessive if it is deployed in Indian waters, offshore Bombay. It is important to calculate annual production downtime due to excessive motions (pitch greater than 2 degrees or roll greater than 12 degrees), if any and take its cognizance while doing production output planning.

C2.6 Offloading Facilities, Shuttle Tankers and Storage

As stated previously, the stabilized crude is continuously offloaded to a tanker for storage and transportation to shore. Marginal oil fields cannot economically justify the capital costs of a transportation pipeline system unless it is short in length, i.e. near an existing pipeline network.

Let's understand the shuttle transportation options and its layout. In the current field development scenario with FPS, three shuttle transportation options, as listed below, have been investigated and the third one is selected for the envisaged purpose:

- A permanently moored storage tanker with a shuttle tanker
- Two loading buoys and two shuttle tankers
- Two shuttle/storage tankers (contract storage/transportation)

Out of these three options, which one to choose depends upon the project economics. The permanently moored storage tanker is the most expensive option of the three as it requires purchasing of the storage tanker and its permanent mooring system besides arranging a shuttle tanker transportation contract. The second option requires two CALM loading buoys and two shuttle tankers and this too normally doesn't workout owing to economic reasons. The third option of contracting with a shipping company to provide storage and shuttle service seems to be economically justified most of the times as it represents the least amount of capital cost expenditure. In this third case, a Catenary Anchor Leg Mooring (CALM) buoy system is purchased and installed in the field and a tanker shipping company will provide storage and shuttle service. In order to enable continued production, the shuttle tanker must be present in the field for 100% of the time. The semi-submersible is designed to accommodate a maximum of eight hours of oil storage to enable tanker change-out at the CALM buoy.

In the present instant, a Catenary Anchor Leg Mooring (CALM) buoy system with a hawser has been installed in water depths to 150 meters and is proposed as the tanker loading system for use with semi-submersibles. The CALM buoy is anchored by six 3-inch ORQ stud-link chain legs, with each leg anchored to a drive pile. The mooring system is capable of surviving the 100 years storm. The buoy is 36 feet in diameter. The proposed shuttle vessel is a 50,000-DWT tanker (375,000 barrels) suitable for transportation of high pour point oil. The tanker is moored to the buoy with a rope hawser. The hawser system and a fluid swivel are arranged on a turntable atop the buoy, to permit tanker weathervaning. The processed oil is pumped directly from the semi-submersible at 15,000 bopd maximum, during normal production operation, to the buoy via 6-inch flexible risers and a pipeline. The risers are arranged in a "Steep S" configuration. The CALM buoy is located 1.5 kilometers from the semi-submersible FPS to avoid interference with tanker movements.

It is assumed that minimal storage (maximum 14,410 barrels in drill-water tanks) onboard the semi-submersible is available and that the production is normally pumped directly to the shuttle tanker. The minimum shuttle tanker size is approximately 17,000 DWT, to contain 5.5 days of production. The number of shuttles required is two, assuming one being loaded, while the other is en-route to shore and back. However, for economic reasons, two 50,000-DWT shuttle tankers shall be used. However, they will not be dedicated to serve the field alone. During shuttle tanker changeover, the production has to be produced temporarily to the onboard tanks. A storage capacity for 8 hours' production (which works out as 5000 barrels or 810 m³) and a loading rate of 1000 bph is proposed.

It is very important to take due cognizance of the various components of downtime that affects offloading. In the present instant, a typical offloading system downtime, estimated at 8%, consisting of the following: Breakdown Maintenance (2%), Major Maintenance and Inspections (2%), Subsea Maintenance (3%) and Offloading Downtime Because of Wave Exceedance (1%).

C2.7 Sub-Sea Template System

For the semi-submersible FPS, normally two sub-sea development options are available:

- Template Wells Development
- Clustered Satellite Wells Development

In most of the cases, however, the template development gets selected for the simple reason that it is available at comparatively lower capital cost and fewer risers required as manifold being at sub-sea.

Components: Sub-Sea Template System

The template is a unitized structure comprised of four principal systems:

- *Template Structure* (Equipment space frame).
- *Template Production System* that comprises of sub-sea trees for well control, Flow lines, Risers and a Manifold to minimize number of risers.
- *Flow lines and Risers.*
- *Control and Data Acquisition System* which enables remote control and monitoring of trees and manifold.

C2.7.1 *Template Structure*

The template structure is a space frame enclosing all sub-sea production equipment inside an open framework of structural members. The structure positions the well slot and manifold to optimize the structural size and configuration. It also provides a guidance system (guideline) for drilling and completion operations. The structure is designed for installation by a crane barge and incorporates a mud mat support and hydraulic leveling system for leveling the structure prior to pilling to the seafloor. The structure has eight well slots (six wells plus two spare slots), four on either side of a central manifold. A pipeline and umbilical connection area is provided on one end of the structure to facilitate diver-assist sub-sea line connections.

The principal dimensions of the structure (for example, in the present case) are:

Width	: 15.2 m (50 ft)
Length	: 30.5 m (100 ft)
Height	: 6.0 m (20 ft)
Weight	: 272 tones (300 tons)
Well Spacing	: 5.2 m (17 ft)

C2.7.2 *Template Production System*

The template production system is comprised of six sub-sea wet trees and a manifold. The trees are identical assemblies, consisting of a master valve block, wellhead connector, flow control choke, and annulus and production piping. The primary purpose of the trees is to provide primary well control and production flow control. The tree assemblies are installed as a module after well completion via a semi-submersible drilling rig. Divers are utilized to make the tree piping connections to the manifold. Just to develop an understanding, the following is a general outline specification of the proposed tree.

- Diver-assist wet trees
- 3 × 2-inch, 5000 psi WP (working pressure)
- Dual-bore mono-block
- Tie-back mandrel for production and annulus bores
- Hydraulic wellhead connector
- Hydraulic fail-safe close valves
- Hydraulic choke assembly
- Valve and choke positions, pressure and temperature indication on annulus and production bores
- Direct hydraulic control
- Independent data acquisition system

The manifold provides the flow path between the template wells and the subsea pipelines. The manifold is integral with the template structure but is designed to be diver maintainable. The manifold combines the production flow lines from all template wells into a 203-millimeter (8-inch ID) production header. A test header is provided to enable individual well production to be diverted to the FPS for well testing. The test header is of the same size and is looped with the production header. The test header is also used for round-trip pigging with the pigs launched and received on the FPS, as well as providing a redundant production flow path to the FPS. An annulus header is provided on the manifold and is tied into all the template wells' annuli. The annulus system enables constant well annulus pressure monitoring and bleed-down. The annulus piping is crossed over to the production piping on the tree assemblies. This arrangement enables flushing and testing of the production piping and flow control choke.

All pipeline and control umbilical connections to the FPS are located at one end of the manifold. Placing the line connections at one end provides a means to optimize line routing to the FPS and provides sufficient working area to permit divers to make up the line connections.

C2.8 Flow Lines and Risers

The flow lines and risers provide the connecting links for the flow path connection between the FPS and the template.

The lines have an approximate 300-meter (1000-foot) length and connect to the template and riser based using divers and diver-assist connection systems. The lines are initiated at the production template and are laid toward the riser bases prior to installation of the risers. Second ends are laid down directly into the riser bases, with length discrepancies accommodated by a sliding "L" spool on the riser base. As such, the length of these lines and positional tolerance of the bases relative to the template are critical. This is justifiable because of the short distance involved. Export lines are initiated at the semi-submersible's riser base and are laid toward the CALM buoy. Second-end connection is by a 180-degree sweep at the riser base for the buoy. Connection methods utilize come-alongs and Chinese fingers for manipulation of the pipe, together with swivel flanges and hydraulic bolt tensioners for make-up.

In general, Flexible flow lines are selected over steel lines for the following reasons:

- Lower installed cost compared with double-insulated, rigid steel pipe.
- Ease of installation and recovery, and can be reused.
- Built-in thermal insulation and electrical heat-tracing.

Here in this case, Flexible production risers are proposed for the FPS for the following reasons:

- The flexible riser system can remain connected for both minor and major well work-over, permitting uninterrupted production operations while performing well work-over, if desired.
- The flexible riser system does not require the rig's mono-pool or draw-works, permitting their use for a second riser system to enable minor and major well work-over.

- The flexible riser system is comprised of a minimum number of components and permits a larger motion envelope for the semi-submersible than a rigid riser system would.
- Minimum deck loads associated with the riser system's static and dynamic loads.

Riser configuration is important. So it is important to undertake a static analysis of various riser configurations using the appropriate software. Here in this case, the "Steep S" riser configuration was selected for all flexible risers because the risers are able to remain connected in the 100 years storm and the risers do not exceed minimum bending radii during all anticipated FPS motions.

For example, in the present case, the principal dimensions of the riser configuration are:

Buoy Uplift	:	50,000 lbs (8 in and 6 in)
		30,000 lbs (3 in)
Buoy Height	:	60 m (200 ft) – equals half of water depth
Length of Catenaries	:	228 m (750 ft)
Standoff Distances	:	80 m (260 ft) (8 in and 6 in)
Buoy-to-Riser Connector		
On FPS at Neutral Position	:	100 m (300 ft) (3 in)

There are four riser bundles, each consisting of three independent lines (viz. two control bundles, one on either side of a single hydrocarbon handling line). The riser bases are installed and anchor piled to the seafloor during the template installation stage. The risers are installed after the FPS is on station, using an appropriate Multi-Service Vessel (MSV) and divers. The risers are attached to the riser bases using diver made-up flange connections, and connect to the FPS by an emergency quick-disconnect hydraulic coupler. The umbilical are one piece and approximately 1300 meters long. They are positioned and supported by the mid-water buoy, to enable a "Steep S" configuration for the umbilical.

A typical outline specification for the risers and flow lines (Production and Annulus lines and also the control Umbilical lines) can be understood from the components-specifications given at **Table 7** at the back of the book.

C2.9 Control and Data Acquisition System

The control system for remote control and monitoring of the sub-sea template wells and manifold is a direct hydraulic control with an independent data acquisition monitoring system. The control equipment located on the FPS includes:

- Hydraulic fluid supply and power unit
- Directional valve control panels (one for each well, plus the manifold)

The FPS sub-sea control equipment is hard-wired into the control room for interface with the overall FPS control system.

The hydraulic control umbilical (six wells, plus one manifold) are totally independent of each other. The data acquisition monitoring system is totally independent of the hydraulic control system. The instrumentation and monitoring data is gathered at a central distribution pod located on the manifold, to facilitate connection to the FPS's electrical umbilical.

C3 PROJECT SCHEDULE

After undertaking all due deliberations and due-diligence over the major project activities as given below, a project schedule is prepared which is separate and unique to a given project or development plan. In the present case under this development option, major project activities are:

(a) Design, fabrication, and offshore installation of template for drilling;
(b) Drilling and completion of six template wells;
(c) Preparation of bid documents for a semi-submersible FPS, and contract award;
(d) Conversion of semi-submersible to an FPS, and fabrication of associated topside facilities;
(e) Installation of the sub-sea Christmas trees and sub-sea manifold to the template;
(f) Installation of risers, flow lines/umbilical, and CALM loading buoy; and
(g) Offshore hookup and commissioning of the semi-submersible floating production system.

While developing the project schedule for this development, following assumptions were made:

(a) Conversion of the semi-submersible rig to an FPS is carried out in an Indian shipyard/south east Asia
(b) The topside facilities are fabricated in a local fabrication yard in India;
(c) The wells are drilled and completed by a separate semi-submersible rig;
(d) Drilling and completion of each well takes about 40 days
(e) The sub-sea Christmas trees and manifold are installed after completion of drilling by the FPS; and
(f) Drilling and production cannot be carried out simultaneously.

Based on the above project activities and assumptions, a project schedule has been drawn indicating that first oil production from the field can begin 14 months after project approval.

C4 PRODUCTION PROFILE

As in the previous two development scenarios, drilling of the wells in this development is assumed to be completed before the tow-out of the semi-submersible FPS. The field will be produced at the maximum rate of 15,000 bopd from the time of the start of production. It will continue at this peak production rate for about 1-3/4 years, after which it will decline gradually to about 3000 bopd at the end of year 5. To allow for production downtime, it was assumed the field will be produced about 330 days per year. An estimated 18 million barrels of oil will be produced at the end of field life.

C5 CAPITAL AND OPERATING COSTS

Now, after understanding all the concepts and attributes related to Semi-submersible based FPS, let's now understand how the project economics gets worked out. I will suggest the students to go through the concept of "Capital Budgeting" and "Project Economics" to understand the feasibility report preparation exercise things in totality. Working out the capital and operating costs is one very important exercise in this direction. While working out these two costs, two scenarios arise:

Case I: Assume Oil Company owns Semi-submersible based FPS

Case II: Assume Oil Company gets Semi-submersible based FPS on Lease

A table given below reflects how the capital and operating costs are worked out and what the various components of these costs are.

The estimated capital cost is about US$ 120.75 million for Case I and about US$ 71.50 million for Case II. The corresponding annual operating costs at peak production rate for Case I and Case II are US$ 9.0 million and US$ 23.7 million respectively.

One very important aspect to consider here is that at which place we are going to undertake the needed (if any) fabrication exercise: locally or at some fabrication yard elsewhere. The contract value against this is worked out accordingly. Here in this case, it is anticipated that fabrication of the sub-sea template and manifold, CALM loading buoy, processing facilities, and rig conversion can be carried out locally for a total contract value of about US$ 13.41 million against these works inclusive of material cost.

One very crucial consideration here is where we are putting our abandon-ment cost. In the present case, the field abandonment cost is included in the contingency for capital cost.

The other important consideration is the assumptions made while working out the leasing rate and the components of the leasing equipment/systems. In the present case, the lease includes the converted semi-submersible, with its added process facilities and utilities at place of deployment, the costs of mobilization, demobilization of the semi-submersible from/to the Gulf of Mexico or the east coast of Canada, and re-conversion back to a drilling rig at the end of the contract period. The leasing rate of US$ 35,655 per day (or US$ 13.015 million per year) for this semi-submersible FPS is calculated based on the following assumptions:

Lease period – 5 years
Lease company – 10%
Interest on dept – 14%
Lease company return on equity – 30%
Salvage value – 80% annually (i.e. 33% after 5 years)

The tables given below reflect how the capital and operating costs are worked out and what the various components of these costs are. The similar exercise in the similar fashion can be worked out in both the other options i.e. field development with FPSO Barge and Tanker. Figures are indicative only.

Semi-Submersible FPS and Template Wells Development
Case I—Assume Oil Company Owns Semi-Submersible Based FPS
Estimated Capital Cost

	US$ (Million)
A. *Semi-submersible FPS*	
(1) Purchase Semi-submersible Rig	40.00
(2) Rig Conversion	1.09
(3) Process Facilities and Utilities	3.32
(4) Offshore Hookup and Commissioning	2.50
Sub Total	46.92
B. *Sub-sea Template Wells (6 wells and Manifold)*	
(1) Template (Incl. Manifold) and Installation	6.00
(2) Drilling and Well Completion	26.28
(3) Sub-sea Trees and Control Systems	7.96
(4) Offshore Installation	3.51
Sub Total	43.75
C. Semi-submersible FPS	
(1) Flow Lines (Assumed Coflexip Pipes)	0.82
(2) Risers (Assumed Coflexip Pipes)	3.81
(3) Offshore Installation	1.88
Sub Total	6.51

D. CALM Loading Buoy and Export Line
    (1) CALM Loading Buoy 5.00
    (2) 6-Inch Diameter Export Line and Riser 1.10
    (3) Offshore Installation 0.69

     Sub Total 6.79

E. Engineering and Project Management
    (1) 10% of Items A to D, but exclude A (1) 6.40
F. Contingency (10%) 10.39

     Total **120.75**

Semi-Submersible FPS and Template Wells Development
Case II—Assume Oil Company gets Semi-Submersible
FPS on Lease
Estimated Capital Cost

	US$ (Million)
A. Semi-submersible FPS	
(1) Tow to Site	0.50
(2) Offshore Hookup and Commissioning	2.00
Sub total	2.50
B. Sub-sea Template Wells (6 wells and Manifold)	
(1) Template (Incl. Manifold) and Installation	6.00
(2) Drilling and Well Completion	26.28
(3) Sub-sea Trees and Control Systems	7.96
(4) Offshore Installation	3.51
Sub total	43.75
C. Flexible Flow Lines and Production Risers	
(1) Flow Lines (Assumed Coflexip Pipes)	0.82
(2) Risers (Assumed Coflexip Pipes)	3.81
(3) Offshore Installation	1.88
Sub total	6.51
D. CALM Loading Buoy and Export Line	
(1) CALM Loading Buoy	5.00
(2) 6-Inch Diameter Export Line and Riser	1.10
(3) Offshore Installation	0.69
Sub Total	6.79
E. Engineering and Project Management	
10% of Items A to D)	5.96
F. Contingency (10%)	5.96
Total	**71.47**

Semi-Submersible FPS and Template Wells Development
Case II—Assume Oil Company Owns Semi-Submersible FPS
Estimated Operating Cost

	US$ (×1000)
1. Semi-submersible FPS Operations and Maintenance	4,380
2. Template Wells Maintenance	200
3. Shuttle and Storage Tanker Operations	1,700
4. Pipeline and Riser Maintenance (2% of capital cost)	130
5. CALM Buoy and Loading Riser Maintenance	170
6. Head Office Support and Administration	1,740
7. Contingency (10%)	815
Total	**$ 8,965 (×1000)**
Say	**$ 9,000 (×1000)**

Semi-Submersible FPS and Template Wells Development
Case II—Assume Oil Company gets Semi-Submersible
FPS on Lease
Estimated Operating Cost

	US$ (×1000)
1. Semi-submersible FPS Lease(US$ 35,655 per day)	$ 13,015
2. Semi-submersible FPS Operations and Maintenance	4,380
3. Template Wells Maintenance	200
4. Shuttle and Storage Tanker Operations	1,700
5. Pipeline and Riser Maintenance	130
6. CALM Buoy and Loading Riser Maintenance	170
7. Head Office Support and Administration	1,740
8. Contingency (10%)	2,133
Total	**$ 23,468 (×1000)**
Say	**$ 23,500 (×1000)**

CONCLUSION AND RECOMMENDATIONS

After undertaking due analysis of all the systems for development, Semi-submersible-based FPS's appears to be well suited for application at offshore India. This FPS concept is well proven in other parts of the world too the one being at Brazil and at Indonesia.

A semi-submersible FPS is well suited for several field applications over its life expectancy because of its ability to provide good stability over a large range of environmental conditions. Also, its mooring system is capable of handling a large range of water depths with minimum or no system modifications. Accordingly, the semi-submersible-based floating production system has several potential applications for use offshore India like for early Production Development, for Marginal Field Developments and also for Extended Well Testing.

There are certain definite advantages of a Semi-Submersible-based FPS compared to a barge-or tanker-based FPS:

- It suits well suited for use at offshore and at several potential sites.
- Most flexible system for reuse with minimum modifications.
- System has the best uptime for continuous production related to weather conditions.
- The semi-submersible rig, as an FPS, is operationally the most versatile vessel when compared to the tanker-or barge-based FPS's. Positioning the sub-sea wells beneath the rig enables well maintenance to be performed without chartering a diving support vessel or drilling rig. It also enables down-hole and wire-line maintenance tasks to be performed from the rig without having to charter a work-over rig. The semi-submersible FPS offers the lowest sub-sea well maintenance costs compared to the other FPS systems, as the system not only saves on charter costs but also minimizes lost production revenues due to immediate access capability.
- When the field is to be abandoned, the rig can recover the sub-sea trees and down-hole completion, and abandon the wells, thus greatly reducing field's abandonment costs.
- Although initial capital costs are high, they are offset against the system's flexibility for reuse over a wide range of water depths with minimum modifications, and overall lower sub-sea maintenance and well abandonment costs.

- The semi-submersible FPS, sub-sea template and wells, and flexible riser systems present no technical concerns, as these systems have been used in several similar applications elsewhere quite successfully.

TO CONCLUDE

Semi-submersible-based FPS is well suited for application offshore India. Several reasons have been attributed to this as we have studied/deliberated in different field development options. In Indian Offshore waters, several small hydrocarbon reservoirs have been found in the water depths ranging from 75 to 200 meters and the depth is within the water depth range of semi-submersibles. Further, Offshore India environmental conditions permit the semi-submersible to remain fully operational in a 10 years storm frequency and to remain on station and fully connected to the sub-sea system in a 100 years storm frequency. Short-term semi-submersible FPS's normally require a tanker storage/shuttle system for crude oil transportation. The CALM buoy system has been used in water depths up to 150 meters, and the relatively calm weather conditions off the coast of India should give favorable system uptime. Tanker shipping companies and tanker shore facility infrastructures are in place at Indian shores, providing an economical means of oil storage and transportation.

Tables

[Figures put here are just an approximations and hypothetical one to comprehend the concept and to develop an understanding]

Table 1
Separator Dimensions

FPS	Description	No.	Design Separation Pressure	Separation Temperature	Separator Dimensions (mm dia × mm)
Barge	HP Separator (V101)	1	1342 kPa (180 psig)	50°C (122°F)	1200 × 4500
	LP Separator (V102)	1	445 kPa (50 psig)	50°C (122°F)	1200 × 4500
	Atm. Separator	1	121 kPa (3 psig)	50°C (122°F)	1200 × 4500
Tanker or Semi-submersible	HP Separator (V101)	1	1342 kPa (180 psig)	50°C (122°F)	1200 × 6000
	LP Separator (V102)	1	445 kPa (50 psig)	50°C (122°F)	1200 × 6000
	Atm. Separator	1	121 kPa (3 psig)	50°C (122°F)	1200 × 6000

Contents of the tables 1 to 10 are compiled from various in-house studies and deliberations undertaken by ONGC.

Table 2
Equipment List—FPSO Tanker Case

Estimated capital costs for additional process facilities, utilities, living quarters, kill facilities plus few miscellaneous works out to be $ 5,885,000 as detailed below.

Equipment Description	Identity No.	Design Conditions Capacity	Pressure	Temp.	W × L × H mm	Oper. Wt. Tonnes	Installed Est. Costs US$
A. Process Facilities							
HP Separator	V-101	20,000 bpd	180	175	1800 dia × 6000	30.0	220,000
LP Separator	V-102	20,000 bpd	50	140	1800 dia × 6000	28.0	200,000
Atm. Separator	V-103	20,000 bpd	5	140	1800 dia × 6000	25.0	185,000
Fuel Gas Scrubber	V-104	11 MMscf/d	180	140	1500 dia × 3000	4.0	50,000
Test Separator	V-105	10,000 bpd	180	175	1800 dia × 6000	30.0	220,000
HP Flare Drum	V-106	11 MMscf/d	50		1500 dia × 3000	4.0	35,000
LP Flare Drum	V-107	4 MMscf/d	20		900 dia × 3000	3.0	25,000
Plate Coalescer	V-108	16,000 bpd			2100 dia × 4500	26.0	150,000
Crude Pumps	P-101 A/B	600 USgpm			900 × 2100 × 600	3.0	40,000
Mol. Sieve	V-110 A/B					15.0	100,000
Condensate Pump	P-102	40 USgpm	60	140	600 × 1200 × 500	1.5	5,500
Recovered Crude Pump	P-103	60 USgpm	60	140	600 × 1200 × 500	2.0	6,500
Inlet Manifold	6-well				4500 × 6000	9.0	150,000
Meter Prover					4500 × 20,000 × 4500	60.0	700,000
Insulation and Heat Tracing						50.0	161,000
Ground Flare		15 MMscf/d				80.0	300,000
Misc. Piping						100.0	–
Misc. Structure and Walkways						60.0	120,000
Instruments						20.0	–
Electrical						20.0	–
Total Process Facilities						**570.5**	**2,668,000**

contd...

Table 2 (contd.)

Equipment Description	Identity No.	Design Conditions				Oper. Wt. Tonnes	Installed Est. Costs US$
		Capacity	Pressure	Temp.	W × L × H mm		
B. Utilities							
Cooling Water Pumps	25-P1	400 USgpm			1200 × 2400 × 900	3.0	11,000
Cooling Water Pumps	25-P2	400 USgpm			1200 × 2400 × 900	3.0	11,000
Cooling Water Surge Vessel	25-VI	5000 USgpm			1800 dia × 7500	21.0	16,000
Piping System						5.0	20,000
Crude Loading Pumps			Existing				
Crude Loading Pumps			Existing				
Stripping Pump	100-P3		Existing				
Stripping Pump	100-P4		Existing				
Piping System						25.0	100,000
Boiler	110-H1		Existing				
Crude Heater	110-E1	80 MM Btu/h			2400 dia × 9000	40.0	500,000
Firewater Pump	17-P1	1000 USgpm	Existing				
Firewater Pump	17-P2	1000 USgpm	Existing				
Firewater Surge Vessel	17-V1	1000 USg	Existing				
Foam Conc. Storage	17-V2	5000 USg			1800 dia × 7500	21.0	9,000
Piping and Other Systems						15.0	60,000
Seawater Pump	24-P1	500 USgpm				2.0	24,000
Seawater Pump	24-P2	500 USgpm				2.0	24,000
Filter	24-S1				600 dia × 1800	1.0	7,000
Seawater Cooler	24-E1				600 dia × 1800	2.5	35,000
Piping System						3.0	50,000
Air Compressor Package	18-K1/K2	100 scfm			1500 × 4800 × 1800	10.0	80,000

Note: (Cost is not considered for existing utility facility); However, if these has to be added, add the cost according)
Note: USgpm: US gallons per minute; Usg: US gallons; scfm: standard cubic feet per minute; bbl: barrels.

contd...

Table 2 (contd.)

Equipment Description	Identity No.	Design Conditions				Oper. Wt. Tonnes	Installed Est. Costs US$
		Capacity	Pressure	Temp.	W × L × H mm		
Instr. Air Dryer	18-M1						
Plant Air Receiver	18-V2						
Instr. Air Receiver	18-V1						
Piping System						7.5	30,000
Diesel Storage Tank	15-D1	2362 bbl			Ships Tank	230.0	
Diesel Transfer Pump	15-P1/P2	15 gpm			600 × 1200 × 600	2.0	12,000
Diesel Filter Separator	15-V1				600 × 600 × 600	1.0	5,000
Piping System						2.5	10,000
Hypochlorite System	23-M1				1800 × 1500 × 1700	2.5	40,000
Inert Gas Generator	20-F1		Existing		1800 × 1500 × 4500	20.0	300,000
Diesel Power Generator	22-G1	500 kW	Existing				
Diesel Power Generator	22-G2	500 kW	Existing				
Diesel Power Generator	22-G3	500 kW	Existing				
Switchgear					600 × 1800 × 3000	2.0	160,000
Rev. OSM Unit incl. Piping	16-M1	100 USgph			1200 × 4500 × 1800	10.0	130,000
Potable Water Tank	16-D1	2250 USg			1800 dia × 3600	10.0	6,500
Potable Water Pump	16-P1	12 USgpm			600 × 1200 × 600	1.0	6,000
Sewage Treatment Plant	16-M3				1800 × 2700 × 2400	35.0	130,000
Chemical Injection Fac.		200 bbl				35.0	60,000
Walkways and Misc. Structure						45.0	
Electrical, etc.						50.0	–
Instruments						50.0	–
Misc. Piping						80.0	–
Total Utilities Systems						**807.0**	**1,836,500**

contd...

Table 2 (contd.)

Equipment Description	Identity No.	Design Conditions				Oper. Wt. Tonnes	Installed Est. Costs US$
		Capacity	Pressure	Temp.	W × L × H mm		
C. Living Quarters							
Living Quarters	40 men	Existing					
Helideck					22,000 × 22,000	290.0	240,000
Helicopter Refuelling Unit		1000 gal			1000 × 1000 × 3800	4.0	21,000
Total Weight—Living Quarters and Helideck						294.0	261,000
D. Kill Facility							
Kill Pumps						20.0	220,000
Total Weight—Kill Facility						20.0	220,000
E. Miscellaneous							
Control Panel						10.0	200,000
Crane (20-tonne) (2 ea.)						40.0	400,000
Survival Crafts (30-men, 2 ea.)	60 men	Existing					
Communication Equipment						5.0	200,000
Battery Charges and Batteries						1.0	50,000
Store						60.0	50,000
Total Miscellaneous						116.0	900,000
Grand Total						**1807.5** **Say 1850 tonnes**	**5,885,000**

Table 3
Equipment List—FPSO Barge Case

Estimated capital costs for additional process facilities, utilities, living quarters, kill facilities plus few miscellaneous works out to be $ 8,213,500 as detailed below.

Equipment Description	Identity No.	Capacity	Design Conditions Pressure	Temp.	W × L × H mm	Oper. Wt. Tonnes	Installed Est. Costs US$
A. Process Facilities							
HP Separator	V-101	10,000 bpd	180	175	1200 dia × 4500	18.0	180,000
LP Separator	V-102	10,000 bpd	50	140	1200 dia × 4500	16.0	75,000
Atm. Separator	V-103	10,000 bpd	5	140	1200 dia × 4500	14.0	50,000
Fuel Gas Scrubber	V-104	4,8 MMscf/d	180	140	900 dia × 3000	2.5	40,000
Test Separator	V-105	10,000 bpd	180	175	1200 dia × 4500	18.0	180,000
HP Flare Drum	V-106	4.8 MMscf/d	50		900 dia × 3000	3.0	25,000
LP Flare Drum	V-107	1 MMscf/d	20		900 dia × 3000	3.0	25,000
Plate Coalescer	V-108	8000 bpd	Atm		1500 dia × 4500	22.0	80,000
Crude Pumps	P-101 A/B	350 USgpm	60		600 × 1800 × 600	2.5	30,000
Mol. Sieve	V-110 A/B					15.0	100,000
Condensate Pump	P-102	30 USgpm	60	140	600 × 1200 × 500	1.5	5,500
Recovered Crude Pump	P-103	40 USgpm	60	140	600 × 1200 × 500	2.0	6,500
Inlet Manifold	4-well				3000 × 4000	9.0	150,000
Meter Prover					4500 × 20,000 × 4500	60.0	655,000
Insulation and Heat Tracing						50.0	161,000
Flare excluding Pipeline							50,000
Misc. Piping						100.0	–
Misc. Structure and Walkways						50.0	100,000
Instruments						20.0	–
Electrical						20.0	–
Total Process Facilities						**426.5**	**1,913,000**

contd....

Table 3 (contd.)

Equipment Description	Identity No.	Design Conditions Capacity	Pressure	Temp.	W × L × H mm	Oper. Wt. Tonnes	Installed Est. Costs US$
B. Utilities							
Cooling Water Pumps	25-P1	350 USgpm			1000 × 1500 × 600	2.0	9,000
Cooling Water Pumps	25-P2	350 USgpm			1000 × 1500 × 600	2.0	9,000
Cooling Water Surge Vessel	25-VI	5000 USgpm			1800 dia × 7500	21.0	16,000
Piping System						5.0	20,000
Crude Loading Pumps	100-P1	3000 bbl/h (2100 Usgpm)			1800 × 3000 × 900	20.0	70,000
Crude Loading Pumps	100-P2	3000 bbl/h (2100 Usgpm)			1800 × 3000 × 900	20.0	70,000
Stripping Pump	100-P3	250 Usgpm			1200 × 1200 × 1800	4.0	30,000
Stripping Pump	100-P4	250 Usgpm			1200 × 1200 × 1800	4.0	30,000
Piping System						25.0	100,000
Boiler	110-H1	50 MM Btu/h			4500 × 4500 × 6000	40.0	1,900,000
Crude Heater	110-E1	40 MM Btu/h			1200 dia × 9000	35.0	400,000
Firewater Pump (Diesel)	17-P1	1000 USgpm			1200 × 2800 × 900	4.0	143,000
Firewater Pump (Electrical)	17-P2	1000 USgpm			1200 × 2800 × 900	4.0	60,000
Firewater Surge Vessel	17-V1	1000 USg			1200 dia × 3300	5.0	4,000
Foam Conc. Storage	17-V2	5000 USg			1800 dia × 7500	21.0	9,000
Piping and Other Systems						15.0	60,000
Seawater Pump	24-P1	500 USgpm				2.0	24,000
Seawater Pump	24-P2	500 USgpm				2.0	24,000
Filter	24-S1				600 dia × 1800	1.0	7,000
Seawater Cooler	24-E1				600 dia × 1800	2.5	35,000
Piping System						3.0	50,000
Air Compressor Package	18-K1/K2	100 scfm			1500 × 4800 × 1800	10.0	80,000
Instr. Air Dryer	18-M1						

contd...

Table 3 (contd.)

Equipment Description	Identity No.	Design Conditions				Oper. Wt. Tonnes	Installed Est. Costs US$
		Capacity	Pressure	Temp.	W × L × H mm		
Plant Air Receiver	18-V2					7.5	30,000
Instr. Air Receiver	18-V1						
Piping System							
Diesel Storage Tank	15-D1	1575 bbl			Barge Tank	200.0	
Diesel Transfer Pump	15-P1/P2	15 gpm			600 × 1200 × 600	2.0	12,00
Diesel Filter Separator	15-V1				600 × 600 × 600	1.0	5,000
Piping System						2.5	10,000
Hypochlorite System	23-M1				1800 × 1500 × 1750	2.5	40,000
Inert Gas Generator	20-F1	0.4 MMscf/d			1800 × 1500 × 4500	20.0	200,000
Diesel Power Generator	22-G1	400 kW			2100 × 6000 × 3000	10.0	200,000
Diesel Power Generator	22-G2	400 kW			2100 × 6000 × 3000	10.0	200,000
Diesel Power Generator	22-G3	400 kW			2100 × 6000 × 3000	10.0	200,000
Switchgear					600 × 1800 × 3000	2.0	200,000
Rev. OSM Unit incl. Piping	16-M1	100 USgph			1200 × 4500 × 1800	10.0	130,000
Potable Water Tank	16-D1	2250 USg			1800 dia × 3600	10.0	6,500
Potable Water Pump	16-P1	12 USgpm			600 × 1200 × 600	1.0	6,000
Sewage Treatment Plant	16-M3				1800 × 2700 × 2400	35.0	130,000
Chemical Injection Fac.		200 bbl				35.0	60,000
Walkways and Misc. Structure						45.0	90,000
Electrical, etc.						50.0	–
Instruments						50.0	–
Misc. Piping						80.0	–
Total Utilities Systems						**831.0**	**4,669,500**

contd...

Table 3 (contd.)

Equipment Description	Identity No.	Design Conditions				Oper. Wt. Tonnes	Installed Est. Costs US$
		Capacity	Pressure	Temp.	W × L × H mm		
C. Living Quarters							
Living Quarters (12" "Porta" Cabins)		24 men			25 × 20 × 3 m	240.02	400,000
Helideck					22 × 22 m	290.0	240,000
Total Weight—Living Quarters and Helideck						530.0	661,000
D. Kill Facility							
Kill Pumps						20.0	220,000
Total Weight—Kill Facility						20.0	220,000
E. Miscellaneous							
Control Panel						10.0	200,000
Crane (20-tonne)						40.0	200,000
Survival Crafts (2 ea.)		20 men				10.0	50,000
Communication Equipment						5.0	200,000
Battery Charges and Batteries						1.0	50,000
Store						60.0	50,000
Total Miscellaneous						**136.0**	**750,000**
Grand Total						**1937.5** Say 1950 tonnes	**8,213,500**

Table 4
Equipment List—Semi-Submersible FPS Case

Estimated capital costs for additional process facilities, utilities, living quarters, kill facilities plus few miscellaneous works out to be $ 3,820,000 as detailed below.

Equipment Description	Identity No.	Design Conditions				Oper. Wt. Tonnes	Installed Est. Costs US$
		Capacity	Pressure	Temp.	W × L × H mm		
A. Process Facilities							
HP Separator	V-101	20,000 bpd	180	175	1800 dia × 6000	30.0	220,000
LP Separator	V-102	20,000 bpd	50	140	1800 dia × 6000	28.0	200,000
Atm. Separator	V-103	20,000 bpd	5	140	1800 dia × 6000	25.0	185,000
Fuel Gas Scrubber	V-104	11 MMscf/d	180	140	1500 dia × 3000	4.0	50,000
Test Separator	V-105	10,000 bpd	180	175	1800 dia × 6000	30.0	220,000
HP Flare Drum	V-106	11 MMscf/d	50		1500 dia × 3000	4.0	35,000
LP Flare Drum	V-107	4 MMscf/d	20		900 dia × 3000	3.0	25,000
Plate Coalescer	V-108	16,000 bpd			2100 dia × 4500	26.0	150,000
Crude Pumps	P-101 A/B	600 USgpm			900 × 2100 × 600	3.0	40,000
Mol. Sieve	V-110 A/B					15.0	100,000
Condensate Pump	P-102	40 USgpm	60	140	600 × 1200 × 500	1.5	5,500
Recovered Crude Pump	P-103	60 USgpm	60	140	600 × 1200 × 500	2.0	6,500
Inlet Manifold	6-well				4500 × 6000	9.0	150,000
Meter Prover					2000 × 15,000 × 2000	20.0	700,000
Insulation and Heat Tracing						50.0	161,000
Flare (2 ea.)						80.0	200,000
Misc. Piping						100.0	–
Misc. Structure and Walkways						30.0	–
Instruments						20.0	–
Electrical						20.0	–
Total Process Facilities						**500.5**	**2448,000**

contd...

Table 4 (contd.)

Equipment Description	Identity No.	Design Conditions				Oper. Wt. Tonnes	Installed Est. Costs US$
		Capacity	Pressure	Temp.	W × L × H mm		
B. Utilities							
Cooling Water Pumps	25-P1		Existing				
Cooling Water Pumps	25-P2		Existing				
Cooling Water Surge Vessel	25-VI		Existing				
Piping System						5.0	20,000
Crude Loading Pumps	100-P1	1000 bbl/h (700 Usgpm)			1500 × 2000 × 6000	4.0	20,000
Crude Loading Pumps	100-P2	1000 bbl/h (700 Usgpm)			1500 × 2000 × 600	4.0	20,000
Stripping Pump	100-P3		Existing				
Stripping Pump	100-P4		Existing				
Piping System						25.0	100,000
Boilers	110-H1/H2		Existing				
Firewater Pump	17-P1	1000 USgpm	Existing				
Firewater Pump	17-P2	1000 USgpm	Existing				
Firewater Surge Vessel	17-V1	1000 USg	Existing				
Foam Conc. Storage	17-V2	5000 USg	Existing				
Piping and Other Systems							60,000
Seawater Pump	24-P1	Existing					
Seawater Pump	24-P2	Existing					
Filter	24-S1				600 dia × 1800	1.0	7,000
Seawater Cooler	24-E1				600 dia × 1800	2.5	35,000
Piping System							50,000
Air Compressor Package	18-K1/K2	100 scfm			1500 × 4800 × 1800	10.0	80,000
Instr. Air Dryer	18-M1						

contd…

Table 4 (contd.)

| Equipment Description | Identity No. | Design Conditions | | | | Oper. Wt. Tonnes | Installed Est. Costs US$ |
		Capacity	Pressure	Temp.	W × L × H mm		
Plant Air Receiver	18-V2						
Instr. Air Receiver	18-V1						
Piping System							30,000
Diesel Storage Tank	15-D1	2362 bbl	Existing		Vessel's Tanks		
Diesel Transfer Pump	15-P1	15 gpm	Existing				
Diesel Filter Separator	15-V1				600 × 600 × 600	1.0	5,000
Piping System							10,000
Hypochlorite System	23-M1				1800 × 1500 × 1700	2.5	40,000
Inert Gas Generator	20-F1	0.5 MMscf/d				20.0	75,000
Diesel Power Generation	22-G1	1500 kW	Existing				
Diesel Power Generation	22-G2	1500 kW	Existing				
Diesel Power Generation	22-G3	1500 kW	Existing				
Switchgear					600 × 1800 × 3000	2.0	160,000
Potable Water Evaporator Unit incl. Piping	16-M1		Existing				
Potable Water Tank	16-D1		Existing				
Potable Water Pump	16-P1		Existing				
Sewage Treatment Plant	16-M3		Existing				
Chemical Injection Fac.		200 bbl				35.0	60,000
Walkways and Misc. Structure						30.0	
Electrical, etc.						40.0	–
Instruments						25.0	–
Misc. Piping						80.0	–
Total Utilities Systems						**262.0**	**872,000**

contd...

Table 4 (contd.)

Equipment Description	Identity No.	Design Conditions				Oper. Wt. Tonnes	Installed Est. Costs US$
		Capacity	Pressure	Temp.	W × L × H mm		
C. Living Quarters							
Living Quarters	40 men		Existing				
Helideck			Existing				
Helicopter Refuelling Unit		1000 gal	Existing				
Total Weight—Living Quarters and Helideck							
D. Kill Facility							
Kill Pumps			Existing				
Total Weight—Kill Facility							
E. Miscellaneous							
Control Panel						10.0	200,000
Crane (30-tonne)			Existing				
Survival Craft		50 men	Existing				
Communication Equipment						5.0	200,000
Battery Charges and Batteries						1.0	50,000
Store						10.0	50,000
Total Miscellaneous						**26.0**	**500,000**
Grand Total						**788.5** Say 790 tonnes	**3,820,000**

Table 5
General Outline Specifications—Flexible Pipes
FPSO Barge and Sub-Sea Wells Development

A. Production and Annulus Lines

Line Size	: 3" ID
Working Pressure	: 2500 psi (172.4 bars)
Design Pressure	: 3855 psi (265.9 bars)
Service	: Sour Crude
	GOR = 560 scf/bbl
	Pour Point = 27°C
	H_2S = 230 ppm
	CO_2 = 3%

General Outline Specifications

- Inner interlocked steel carcass
  - AISI Grade 304 equivalent stainless steel
  - 0.80 mm thick
- Internal pressure plastic sheath
  - Polyamide II Rilsan BENSO P40 TL material
  - 6.0 mm thick
- Zeta spiral (pressure armour)
  - Low carbon steel (conform to NACE's MR 01-75 standards, or equivalent)
  - 4.8 mm thick
- Intermediate (or antifriction) plastic sheath
  - Rilsan material
  - 2.0 mm thick
- Tensile armour
  - Low carbon steel (conform to NACE's MR 01-75, or equivalent)
  - 2 × 1.6 mm thick
- Thermal insulation sheath
  - Coflexip's patented material
- External thermoplastic sheath
  - Rilsan
  - 5.0 mm thick
- Coflexip's patented electric cable system

B. Control Umbilical Lines

Hose Size	: 6 × ¼" ID and 3 × 3/8" ID
Working Pressure	: 5000 psig (345 bars)
Service	: Hydraulic fluid or chemical

General Outline Specifications

- 6 × 1/4" ID and 3 × 3/8" ID hoses with blank fillers
- Intermediate Rilsan sheath
- Interlocked steel carcass conforming to NACE's MR 01-75 requirements, or equivalent
- Intermediate polyethylene sheath
- Tensile armours
- External Rilsan sheath

Table 6
General Outline Specification—Flexible Pipes
FPSO Tanker and Sub-Sea Well Development

A. Production and Annulus Lines

Line Size	:	3" ID
Working Pressure	:	2500 psi (172.4 bars)
Design Pressure	:	3855 psi (265.9 bars)
Service	:	Sour Crude
		GOR = 560 scf/bbl
		Pour Point = 27°C
		H_2S = 230 ppm
		CO_2 = 3%

General Outline Specifications

- Inner interlocked steel carcass
  - AISI Grade 304 equivalent stainless steel
  - 0.80 mm thick
- Internal pressure plastic sheath
  - Polyamide II Rilsan BENSO P40 TL material
  - 6.0 mm thick
- Zeta spiral (pressure armour)
  - Low carbon steel (conform to NACE's MR 01-75 standards, or equivalent)
  - 4.8 mm thick
- Intermediate (or antifriction) plastic sheath
  - Rilsan material
  - 2.0 mm thick
- Tensile armour
  - Low carbon steel (conform to NACE's MR 01-75, or equivalent)
  - 2 × 1.6 mm thick
- Thermal insulation sheath
  - Coflexip's patented material
  - 0.80 mm thick
- External thermoplastic sheath
  - Rilsan
  - 5.0 mm
- Coflexip's patented electric cable system

B. Control Umbilical Lines

Hose Size	:	6 × 1/4" ID and 3 × 3/8" ID
Working Pressure	:	5000 psig (345 bars)
Service	:	Hydraulic fluid or chemical

General Outline Specifications

- 6 × 1/4" ID and 3 × 3/8" ID hoses with blank fillers
- Intermediate Rilsan sheath
- Interlocked steel carcass conforming to NACE's MR 01-75 requirements, or equivalent
- Intermediate polyethylene sheath
- Tensile armours
- External Rilsan sheath

Table 7
General Outline Specifications—Flexible Pipes
Semi-Submersible FPS and Template Wells Development

A. Production and Annulus Lines

Line Size	:	Production 8" ID, Annulus 3" ID
Working Pressure	:	2500 psi (172.4 bars)
Design Pressure	:	3855 psi (265.9 bars)
Service	:	Sour Crude
		GOR = 560 scf/bbl
		Pour Point = 27°C
		H_2S = 230 ppm
		CO_2 = 3%

General Outline Specifications

- Inner interlocked steel carcass
  - AISI Grade 304 equivalent stainless steel
  - 0.80 mm thick
- Internal pressure plastic sheath
  - Polyamide II Rilsan BENSO 240 TL material
  - 6.0 mm thick
- Zeta spiral (pressure armour)
  - Low carbon steel (conform to NACE's MR 01-75 standards, or equivalent)
  - 2 × 1.6 mm thick
- Intermediate (or antifriction) plastic sheath
  - Rilsan material
  - 2.0 mm thick
- Tensile armour
  - Low carbon steel (conform to NACE's MR 01-75, or equivalent)
  - 2 × 1.6 mm thick
- Thermal insulation sheath
  - Coflexip's patented material
- External thermoplastic sheath
  - Rilsan
  - 5.0 mm
- Coflexip's patented electric cable system

B. Control Umbilical Lines

Hose Size	:	9 × ¼" ID and 3 × ³/₈" ID hydraulic lines
Working Pressure	:	5000 psig (345 bars)
Service	:	Hydraulic fluid or chemical

General Outline Specifications

- 9 × ¼" ID and 3 × ³/₈" ID hoses with blank fillers
- Intermediate Rilsan sheath
- Interlocked steel carcass conforming to NACE's MR 01-75 requirements, or equivalent
- Intermediate polyethylene sheath
- Tensile armours
- External Rilsan sheath

Table 8
Typical Power Requirements
for 10,000 bpd Production Rate

BARGE CONCEPT

Equipments	Power	Requirement
Crude Oil Pumps	25 hp	18.75 kW
Oily Water Pump	5 hp	3.75 kW
Coalescer Oil Pump	3 hp	2.20 kW
LP Knockout Drum Pump	3 hp	2.20 kW
Boiler Feedwater Pump	30 hp	22.00 kW
Desalination Unit	25 hp	18.75 kW
Potable Water Pump	5 hp	4.00 kW
Instrument Air Compressor	25 hp	18.75 kW
Diesel Fuel Pump	3 hp	22.00 kW
Seawater Pumps	130 hp	100.00 kW
Jockey Pump	15 hp	11.25 kW
Cooling Medium Pump	15 hp	11.25 kW
Stripping Pumps	40 hp	37.50 kW
Living Quarters		75.00 kW
Battery Charges		10.00 kW
Lighting		50.00 kW
Heating (Misc.)		75.00 kW
Miscellaneous		37.00 kw
	approx.	500.00 kW
Inrush Allowances		300.00 kW
Total		**800.00 kW**

Table 9
Typical Power Requirements
For 20,000 bpd Production Rate

TANKER AND SEMI-SUBMERSIBLE CONCEPTS

Equipments	Power	Requirement
Crude Oil Pumps	25 hp	30.00 kW
Oily Water Pump	5 hp	4.00 kW
Coalescer Oil Pump	3 hp	2.20 kW
LP Knockout Drum Pump	3 hp	2.20 kW
Boiler Feedwater Pump	50 hp	37.50 kW
Desalination Unit	25 hp	18.75 kW
Potable Water Pump	5 hp	3.75 kW
Instrument Air Compressor	25 hp	18.75 kW
Diesel Fuel Pump	3 hp	2.20 kW
Seawater Pumps	160 hp	120.00 kW
Jockey Pump	15 hp	11.25 kW
Cooling Medium Pump	15 hp	11.25 kW
Stripping Pumps	50 hp	37.50 kW
Living Quarters		75.00 kW
Battery Charges		10.00 kW
Lighting		50.00 kW
Heating		75.00 kW
Miscellaneous		90.65 kW
	approx.	600.00 kW
Inrush Allowances		400.00 kW
Total		**1000.00 kW**

Table 10
Storage Design Data for FPS

	FPS Barge Case	FPS Tanker Case	FPS Semi-submersible Case
Field and Production Data			
Route Length of Shuttle Vessel	100 km	260 km	160 km
Shuttle Speed	10 knots	10 knots	10 knots
Production Rate	10,000 bpd	20,000 bpd	20,000 bpd
Offshore Loading Rate	3,000 bph	16,000 bph	1,000 bph
Travel Time of Shuttle	6 hrs	10 hrs	10 hrs
	Say (0.5 day)	(0.5 day)	(0.5 day)
Operational Parameters			
Time Waiting to Moor and Start	0.5 day	0.5 day	0.5 day
Unmooring—Onshore and Offshore	0.5 day	0.5 day	0.5 day
Shore Unloading Time	1.0 day	1.0 day	1.0 day
Offshore Loading Time	1.0 day	1.0 day	1.0 day
Weather and Scheduling Delay	1.0 day	2.0 days	2.0 days
Round-trip Time	4.5 days	5.5 days	5.5 days
Offshore Storage Max. Fill	95%	95%	95%
Offshore Storage Max. Heel	5%	5%	5%
Shuttle Max. Fill	95%	95%	95%
Shuttle Max. Heel	5%	5%	5%

Table 11
Toxicity Level of H₂S with Respect to Other Gases

Chemical Formula	Specific Gravity	Threshold Limit	Hazard Limit	Lethal Concentration
HCN	0.94	10 ppm	150 ppm/hr	300 ppm
H$_2$S	1.19	20 ppm	250 ppm/hr	700 ppm
SO$_2$	2.21	5 ppm	–	1000 ppm
Cl$_2$	2.45	1 ppm	4 ppm/hr	1000 ppm
CO	0.97	50 ppm	400 ppm/hr	1000 ppm
CO$_2$	1.52	5000 ppm	5%	10%
CH$_4$	0.55	90000 ppm	Combustible above 5% in air	

Threshold Limit—concentration at which it is believed that all workers may be repeatedly exposed day after day without adverse effects.

Physical Effects of H₂S

Concentration (ppm)	Physical Effects
02	Odor Threshold
10	Obvious & unpleasant odor
20	Safe for 6 hours exposure

Wear Respiratory Protection Over 20 ppm

Concentration (ppm)	Physical Effects
100	Kills smell in 30 to 15 minutes, may sting eyes and throat
200	Kills smell immediately, stings eyes & throat
500	Dizziness, breathing ceases in a few minutes, needs prompt artificial respiration
700	Unconscious quickly, death will result if not rescued promptly
1000	Unconscious at once, followed by death within minutes

Appendices

APPENDIX A
Process Terminology Normally Used

Term	Meaning
ABP	After Bean Pressure
BDV	Blow Down Valve
DG	Diesel Generator
ESD	Emergency Shutdown
FCV	Flow Control Valve
FE	Flow Element
FSD	Fire Shutdown
FTHP	Flowing Tubing Head Pressure
GIP	Gas Injection Pressure
IA	Insulation Acoustic
IA	Instrument Air
IE	Insulation Electric
ILSL	Interface Level Switch Low
ILSLL	Interface Level Switch Low Low
IT	Insulation Thermal
LAL	Level Alarm Low
LALL	Level Alarm Low Low
LG	Level Gauge
MOL	Main Oil Line Pump
PAH	Pressure Alarm High
PALL	Pressure Alarm Low Low
PCV	Pressure Control Valve
PFD	Process Flow Diagram
PI	Pressure Indicator
PIC	Pressure Indicator Controller
PRV	Pressure Relive Valve
PSH	Pressure Switch High
PSHH	Pressure Switch High
PSHL	Pressure Switch High Low
PSL	Pressure Switch Low

PSLL	Pressure Switch Low Low
PSV	Pressure Safety Valve
RPMC	Remote Panel Monitoring and Control
RTU	Remote Telemetry Unit
SCADA	Supervisory Control and Data Acquisition
SSSV	Sub Surface Safety Valve
SSV	Surface Safety Valve
STHP	Shut-in Tubing Head Pressure
TCV	Temperature Control Valve
TG	Turbine Generator
TI	Temperature Indicator
TIC	Temperature Indicator Controller.
TSH	Temperature Switch High
TSHH	Temperature Switch High High
UG	Utility Generator
WM	Water Maker
XSDV/SDV	Shutdown Valve

APPENDIX B

Radio Alphabets

Language	A	B	C	D	E	F	G
International	**Alpha**	**Bravo**	**Charlie**	**Delta**	**Echo**	**Foxtrot**	**Golf**
British	Andrew	Benjamin	Charlie	David	Edward	Frederick	George
American	Abel	Baker	Charlie	Dog	Easy	Fox	George

Language	H	I	J	K	L	M	N
International	**Hotel**	**India**	**Juliet**	**Kilo**	**Lima**	**Mike**	**November**
British	Harry	Isaac	Jack	King	Lucy	Mary	Nelly
American	How	Item	Jig	King	Love	Mike	Nan

Language	O	P	Q	R	S	T	U
International	**Oscar**	**Papa**	**Quebec**	**Romeo**	**Sierra**	**Tango**	**Uniform**
British	Oliver	Peter	Queenie	Robert	Sugar	Tommy	Uncle
American	Oboe	Peter	Queen	Roger	Sugar	Tare	Uncle

Language	V	W	X	Y	Z		
International	**Victor**	**Whiskey**	**X-ray**	**Yankee**	**Zulu**		
British	Victor	William	X-mas	Yellow	Zebra		
American	Victor	William	X (Eks)	Yoke	Zebra		

APPENDIX C

Fire and Fire Classification

Fire is an uncontrolled exothermic chemical reaction in which air/oxygen, inflammable material and heat energy are subjected beyond ignition temperature. This means three elements, as mentioned below, are involved in occurrence of fire, thereby constituting a "Fire Triangle".

- Fuel
- Air/oxygen
- Heat

Fire triangle

Classification of Fire

Fire is classified into four types:

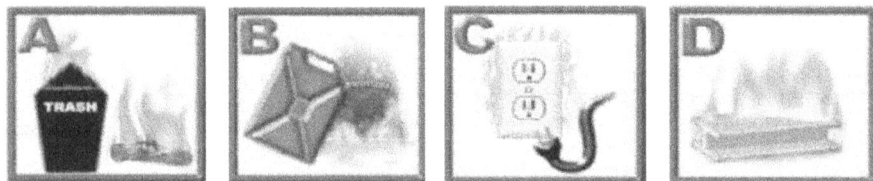

Classification of fire

'Class–A' Fire: These fires are due to ordinary combustible materials such as wood, cloth, paper and plastics. Examples of such materials commonly used in oil field are: Constructive materials and wood decking scaffolding, fiber ropes, clearing rags, tarpaulin etc.

'Class–B' Fire: These are the fires flammable liquids, gases and greases such as oil & condensate, residue from stored hydrocarbons, welding and cutting gases, paints, lubricating and hydraulic fluids etc.

'Class–C' Fire: These are the fires, which involve energized electrical equipments where electrical non-conductivity of extinguisher is of importance. When electrical equipment is de-energized, fire becomes class-A or B fire.

'Class–D' Fire: These are the files of combustible metals such as magnesium, zirconium, sodium and potassium.

APPENDIX D

Classification of Hazardous Area

In order to determine the type of electrical installation suitable for use in different hazardous atmosphere, the hazardous areas have been classified into three zones viz. zone-0, zone-1 and zone-2, according to the degree of probability of the presence of hazardous atmosphere.

- *Zone-0 hazardous area* means an area in which hazardous atmosphere is continuously present and any arc or spark resulting from failure of electrical apparatus in such an area would almost certainly lead to fire or explosion.
- *Zone-1 hazardous area* means an area in which a hazardous atmosphere is likely to occur under normal operating conditions. Such conditions are likely to occur at any time at oil and gas wells and production facilities, which therefore require the most practicable electrical protection.
- *Zone-2 hazardous area* means an area in which a hazardous atmosphere is likely to occur only under abnormal operating conditions, which may be caused only by the simultaneous occurrence of a spark resulting from an electrical failure and a hazardous atmosphere arising through failure of control system.

An electrical spark near a flammable substance can cause serious fire hazard. A fire or explosion can occur if a flammable atmosphere and a source of ignition (spark) exist together. To avoid this, special electrical equipment is to be used depending on the area. In zone-0 area, it is preferable to avoid all electrical equipment; otherwise only intrinsically safe type equipment can be used. In case of zone-1, only "certified flameproof" electrical equipment can be used. If the area comes under the category of zone-2, "non sparking" types electrical equipment can be used.

The electronic control signals to the field instruments are controlled such that the power in the loop is not sufficient to initiate an ignition. All floating production systems are having independent flares, an essential part of the system. These flares are installed at safe distance to avoid any hazard which may be caused due to accumulation of combustible gases on the prevalent wind direction, such that in case the flame gets extinguished than the hydrocarbon gases should not engulf the FPS.

Appendix E

H₂S Hazards and Characteristics

The Principal Hazard of H_2S is DEATH by inhalation. When the amount of gas absorbed into the blood stream exceeds that which is readily oxidized, systemic poisoning results with a general action on the nervous system. Labored respiration occurs shortly and respiratory paralysis will follow immediately at higher concentrations. Death will occur from asphyxiation unless the exposed person is removed immediately to fresh air and breathing stimulated by artificial respiration. Other level of exposure may cause the following symptoms individually or in combination: Headache, dizziness, nausea, cough, drowsiness, eye and throat irritation, dryness and pain in nose, throat and chest. Detection of H_2S solely by smell is highly dangerous as the sense of smell is rapidly paralyzed by the gas.

Physical Effects of H_2S

Concentration (ppm)	Physical Effects
02	Odor Threshold
10	Obvious and unpleasant odor
20	Safe for 6 hours exposure

Wear Respiratory Protection over 20 ppm

Concentration (ppm)	Physical Effects
100	Kills smell in 30 to 15 minutes, may sting eyes and throat
200	Kills smell immediately, stings eyes and throat
500	Dizziness, breathing ceases in a few minutes, needs prompt artificial respiration
700	Unconscious quickly, death will result if not rescued promptly
1000	Unconscious at once, followed by death within minutes

H₂S Characteristics

Following are the characteristics of H$_2$S:

- Extremely toxic, rank second to hydrogen cyanide.
- Colorless and has offensive odor of rotten eggs.
- Heavier than air – sp. gr. 1.19 (air = 1.00 at 60 deg F).
- Burns with blue flame and form an explosive mixture with air in concentration between 4.3 and 46 percent by volume.
- Auto ignition point is 500 deg F (cigarette burns at 1400 deg F).
- Soluble both in water and liquid hydrocarbons.
- Produces irritation to eyes, throat and respiratory system.
- Corrosive to all electrochemical series metals.

Toxicity level of H$_2$S with respect to other gases

Chemical Formula	Specific Gravity	Threshold Limit	Hazard Limit	Lethal Concentration
HCN	0.94	10 ppm	150 ppm/hr	300 ppm
H$_2$S	1.19	20 ppm	250 ppm/hr	700 ppm
SO$_2$	2.21	5 ppm	–	1000 ppm
Cl$_2$	2.45	1 ppm	4 ppm/hr	1000 ppm
CO	0.97	50 ppm	400 ppm/hr	1000 ppm
CO$_2$	1.52	5000 ppm	5%	10%
CH$_4$	0.55	90000 ppm	Combustible above 5% in air	

APPENDIX F

Effects of H₂S on Metals

The source of hydrogen sulfide in nature are varied like thermal alteration of organic sulfur compounds in reservoir, thermo catalytic reduction of sulfate in formation water in contact with reservoir hydrocarbon, decay of microorganism where hydrogen sulphide is released, so on and so forth. National Association of Corrosion Engineers (NACE) specifies 0.05-psia partial pressure of H_2S in the gas phase, a sour system. Hydrogen Sulfide is very corrosive to all electrochemical series metals. It can cause hydrogen embitterment in steel pipe having tensile strength of 95000 psi or more. Corrosion by H_2S includes in general weight loss in the presence of H_2S.

Water is essential for H_2S to be corrosive. In the presence of water, the reaction of H_2S is as follows:

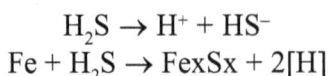

$$H_2S \rightarrow H^+ + HS^-$$
$$Fe + H_2S \rightarrow FexSx + 2[H]$$

With iron, H_2S reacts to form iron sulfide, a black powder scale which adheres to surface as a protective layer. When this coating is not complete, the corrosion continues unabated because iron sulfide becomes cathodic to iron.

The most area of concern is the phenomenon of embitterment and cracking due to penetration of nascent Hydrogen atom into steel at ambient temperature referred to as Sulfide Stress Corrosion, which is more severe than general weight loss corrosion.

In Sulfide Stress Corrosion (SSC), the hydrogen atoms produced as a result of above reactions combine to form molecular hydrogen. As this reaction is retarded, due to the presence of sulfide ions, there is always a fixed concentration of hydrogen atoms present at the metal surface. These hydrogen atoms diffuse into the steel lattice and make it brittle. So when the material is under stress, due to loss of ductility, it fails without being plastically deformed. This failure is sudden and catastrophic and no prior warning is usually indicated. This is known as Sulfide Stress Corrosion. The factors which influence the mechanism of SSC are the characteristics of steel, its structure, mechanical and chemical properties of steel, pH of the fluid (the effect is more when pH is less than 5), fluid temperature (as embrittlement is maximum between 20 to 30 deg C and slight above 80 deg C).

Material Selection for Use in H₂S environment

Metal component used in H_2S service or potential H_2S areas should be those manufactured to resist sulfide stress cracking. NACE MR-01-75 (1984) is a standard for material requirement for sulfide stress cracking resistant metallic materials for oil field equipments. NACE TM 01–077 is a recommended test procedure for selecting material for sour services, the above two standards give the detailed description of material to be used for sour service for different operating conditions (mainly temperature & pressure considerations) and various applications (piping, tunings, casing and accessories there of), NACE Carbon Steel (CS) and Duplex Stainless Steel (DSS).

The NACE carbon steel is the carbon steel manufactured in accordance to NACE standards to meet recommended chemical composition and mechanical properties. The term duplex stainless steel has come to mean a grade whose annealed structure is typically about equal parts of austenite and ferrite. The DSS offer, several advantage over common austenite stainless steels. The DSS are highly resistant to chloride stress corrosion and have excellent pitting and crevice resistance. Further, the use of nitrogen as an alloy addition to DSS allows DSS for use in H_2S environment over 1200 ppm.

APPENDIX G

Desulphurization Process

Desulphurization is a process of removing H_2S from the gas to be utilized. It forms a very important part if we are working in a sour gas field. Hence, it is imperative to understand in some detail this desulphurization process.

There are three desulphurization processes:

- Amine-absorption Process
- Iron-Sponge Process
- Molecular Sieve Process

Amine-Absorption Process

The most common way to remove H_2S and CO_2 from a gas stream is using an amine to absorb the H_2S and CO_2. The Amine-absorption desulphurization process is quite similar to that of glycol dehydration and regeneration system. The undesirable components (H_2S and CO_2) are removed from the gas stream by the amine the absorber, and a regeneration system is utilized to discharge the undesirable components from the circulating solution.

In this process, a water and amine solution that may vary from about 10% to 20% amine is utilized for removing the H_2S and CO_2 from the incoming gas stream. The principle of this process is based on the fact that acid gases (hydrogen sulfide and carbon dioxide) will react with amine at ordinary temperatures, but the reaction is reversed in the regenerator at temperatures of about 220°F to 240°F. The sour gas passes up through the absorber, contacting the lean amine solution passing down the absorber. The foul solution is discharged from the bottom of the absorber and passes through a heat exchanger before it discharges into the top of the still column. The amine solution is boiler in the amine re-boiler by the application of heat. This boiling action supplies steam vapors that pass up through the still column, sweeping the hydrogen sulfide and carbon dioxide from the incoming solution.

The regenerated amine leaving the re-boiler then passes through the amine to the amine heat exchanger, then to the amine storage tank, and is re-circulated with the amine pump. The hydrogen sulfide and carbon dioxide leaving the top of the still column have a considerable volume of steam with them. In order to keep down the quantity of make-up water required, this overhead product is usually cooled to about 120°F. This allows most of the steam to condense so that it can be returned to the amine solution system.

Three types of amine may be used in this process: Mono-ethanolamine, Di-ethanolamine, and Tri-ethanolamine. Mono-ethanolamine (MEA) is preferred in most operations. This is because it is a much stronger base, having a higher affinity to the acid gases. It also has a lower molecular weight, resulting in smaller quantities of amine to be circulated per unit volume of acid gas removed. A "rule of thumb" comparison between MEA and DEA is that a given unit requires approximately 67 times more DEA and MEA for removing the same volume of acid gases. Tri-ethanolamine requires a greater circulation rate than the DEA. One factor affecting the choice of DEA over MEA is that MEA is reputed to react with carbonyl sulfide (COS). This gaseous compound is often present in flue gases and refinery gases. The result of this reaction is unregenerative compounds which tie up with the amine. This reduces the capacity of the solution for removing the acid gases in the absorber.

Iron-Sponge Process

This process utilizes a hydrated iron oxide which reacts only with H_2S. It finds application only on gases with low concentrations of H_2S. The process is intermittent because the iron sulfide must be regenerated or replaced. Care need to be taken during recharging to prevent pollution and/or fire.

The iron-sponge method is a batch process that uses a chemical mechanism for removal of hydrogen sulfide (not carbon dioxide). In the process, the gas to be treated is contacted with a moist bed of wood shavings coated with hydrated iron oxide. The hydrogen sulfide reacts with the iron oxide forming iron sulfide. Air is used to reconvert the iron sulfide to iron oxide, freeing elemental sulphur which deposits on the wood shavings. Eventually, the bed losses its activity by becoming plugged with elemental sulphur and has to be replaced. Because the beds can be only partially regenerated, this process is not considered to be a regenerative process. Also, because the beds have to be replaced periodically, the iron-sponge method is a batch process. The process can be used for removing hydrogen sulfide from gas containing up to 1000 grains/100 scf (0.023 mg/m³). However, it is normally used on streams containing less than 100 grains/100 scf (0.0023 mg/m³).

Molecular Sieve Process

The H_2S concentration is less than 1.5%, the H_2S can be effectively removed with a molecular sieve unit, which is a simple process. Regeneration can be achieved using exhaust gas of temperatures over 550°F, which is available from the steam boiler. Cooling of the bed can be achieved using the inert gas from the inert gas generators. The H_2S/regeneration gas mixture will be disposed of through the flare system. The molecular sieve also removes the water in the gas.

The liquid-free gas leaves the separator, passing down through the active bed in absorber No. 2 where water and H_2S are absorbed on the desiccant. Stripped, dry gas leaves the tower and flows to the outlet lien. While absorber No. 2 is on stream, absorber No. 1 is being regenerated. In the regeneration cycle, a blower circulates regeneration gas. Cool gas from the inert gas generator is passed through absorber No. 1 after regeneration, during the cooling portion of the cycle. During the heating portion of the cycle, hot gas enters the saturated bed, diving off the absorbed water and H_2S. Hot regeneration gas and products from the bed leave the absorber. Final cooling and condensing of the water and hydrocarbons takes place in the regeneration cooling cycle.

At the end of the heating and cooling period, a time-cycle controller switches the raw, untreated gas stream into the regenerated absorber No. 1 and hot regeneration gas into the saturated tower.

APPENDIX H

Corrosion Mechanism

The corrosion mechanisms in gas and oil production are usually divided into four main types:

- Electrochemical corrosion
- Oxygen corrosion
- Sour corrosion
- Sweet corrosion

Out of these four, sour corrosion and sweet corrosion represent the highest risks. So let's understand this in little more details:

Sour Corrosion

Sour corrosion occurs even when very small quantities of H_2S are produced. The chemical reaction showing hydrogen sulfide corrosion is:

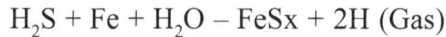

$$H_2S + Fe + H_2O - FeSx + 2H \text{ (Gas)}$$

Hydrogen embrittlement can occur in an H_2S (sour) system. The mechanism occurs when part of the hydrogen produced at the cathode enters the metal substructure instead of going in solution. The hydrogen enters the steel as atomic hydrogen, later to become molecular hydrogen, exerting tremendous pressures on the internal metal structure. This reduces the strength of the steel and, with applied stresses, the steel fails. Hydrogen embrittlement occurs only in high-strength steel (tensile strength 110,000 psi, 78 kg/mm^2).

Steels with hardness above Rc 22 are susceptible to hydrogen stress-cracking below a temperature of 150°F (65°C). Removing water from the system would eliminate the problems of stress-cracking with it. However, contrary to a carbon dioxide corrosion problem, where no disaster would happen if anti-corrosion measures would slip for a week per year, the time within which stress-cracking can occur is a matter of minutes. As treatment cannot be 100%, the material must be crack resistant.

Materials suitable for H_2S service are low-strength materials or materials with a high strength but heat-treated to keep their hardness below a critical level. Welding of materials to be used in sour service should be done with much care to avoid high hardness.

Sweet Corrosion

This form of corrosion results from the presence of carbon dioxide and fatty acids, with no oxygen or H_2S present. The chemical reaction of sweet corrosion is:

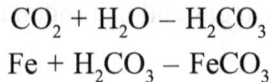

$$CO_2 + H_2O - H_2CO_3$$
$$Fe + H_2CO_3 - FeCO_3$$

CO_2 corrosion is a function of the partial pressure of CO_2. For example, corrosion rates can reach a level of 0.5 mm per year at a partial pressure of 7 psi CO_2 and at ambient conditions.

Corrosion Prevention

Protection from corrosion is accomplished through one of the following methods:

- Conditioning of process stream
    - Inhibition (chemicals, etc.)
      Corrosion prevention by "Inhibition", i.e., conditioning of process stream by use of chemicals
    - Drying (water removal)
      Corrosion prevention by "drying", i.e., conditioning of process streams by water removal
    - Gas removal (stripping)
      Corrosion prevention by "Stripping", i.e., conditioning of process streams by gas removal
- Material selection
- Coatings
- Cathodic protection

NACE Standard MR-01-75 recommends selection of materials which are resistant to sulphide stress-cracking for a partial pressure of H_2S greater than 0.05 psia, if the gas being handled is at a total pressure of 65 psia or greater. If H_2S is less than this, carbon steel can be used for piping and equipment. Further, a 10% variation upward in partial pressure of H_2S makes the piping and equipment subject to sulphide stress-cracking, hence it is important to ensure that the H_2S partial pressure is below 0.05 psia. The CO_2 dissolved in water is not as corrosive as when CO_2 is also present. If CO_2 is not expected in crude, it is recommended to select carbon steel for piping and equipment. Further, we should go for the installation of corrosion inhibitor injection, to cover the risk of corrosion by carbonic acid.

List of Some Useful Sites, Journals

Technology in E&P sector is changing fast. It is quite imperative, hence, to keep tab on the latest developments in the area of floating production systems through internet search, through established journals and con-current books.

List of some useful sites are as follows; however not limited to;

www.fpso.net

www.api.org

www.dnv.org

www.spe.org

www.otcnet.com

www.offshore-technology.com

www.omae.org

www.sname.org

www.isope.org

www.asce.org

www.eagle.org

www.coe.berkeley.edu/issc

www.shipstructure.org

http://ittc.sname.org

List of some useful journals are:

Journal of Petroleum Technology

Oil-Asia

World-Oil

Proceedings of:

1. WPC (World Petroleum Congress)
2. OTC (Offshore Technology Conference)
3. Petro-tech, IORS, India
4. SPE Conferences

Glossary

Some Oil and Gas Fields Terminology

Term	Meaning
AGA	American Gas Association
API	American Petroleum Institute
ASTM	American Society for Testing Material which establishes many of the technical standards used in the oil industry
Abandoned Well	A well not in use because it was a dry hole originally, or because it has ceased to produce.
Abandon	To cease work on a well which is non-productive, to plug off the well with cement plugs and salvage all recoverable equipment Also used in the context of field abandonment.
Acreage	An area, measured in acres, which is subject to ownership or control by those holding total or fractional shares of working interests.
API gravity	The industry standard method of expressing specific gravity of crude oils. Higher API gravities mean lower specific gravity and lighter oils. The measuring scale is calibrated in terms of degrees API; it is calculated as follows: Degrees API $= (141.5/ \text{sp.gr.}60 \text{ deg.F}/60 \text{ deg.F}) - 131.5$
Annulus	The space between the drill string and the well wall or between casing strings or between the casing and the production tubing.
Appraisal Well	A well drilled as part of an appraisal-drilling programme which is carried out to determine the physical extent, reserves and likely production rate of a field.
Associated Gas	Natural gas associated with oil accumulations, which may be dissolved in the oil at reservoir conditions or may form a cap of free gas above the oil.
bcf	The abbreviation for billion cubic feet.
Barrel	A measurement used in the oil industry for a unit of volume of oil or oil products equivalent to 158.978 litres or 42 US gallons.
Barrels per day	A unit of measurement used in the industry for the production rates of oil fields, pipelines, and transportation. Abbreviated to "bpd", "b/d" or "bbl/d".

Barrels of Oil Equivalent (BOE)	Gas volume that is expressed in terms of its energy equivalent in barrels of oil. 6,000 cubic feet of gas equals 1 Barrel of Oil Equivalent (BOE); or 42 US gallons of oil at 40 degrees Fahrenheit.
Blow-Out Preventers (BOPs)	High-pressure wellhead valves, designed to shut off the uncontrolled flow of hydrocarbons. The equipment installed at the wellhead to control pressures in the annular space between the casing and drill pipe or tubing during drilling, completion, and work over operations.
Blow-Out	When well pressure exceeds the ability of the wellhead valves to control it. Oil and gas "blow wild" at the surface. It the uncontrolled flow of gas, oil or other fluids from a well.
Borehole	The hole as drilled by the drill bit.
British Thermal Unit (BTU)	The quantity of heat required to raise the temperature of 1 pound of liquid water by 1 degree Fahrenheit at the temperature at which water has its greatest density (approximately 39 degrees Fahrenheit).
CAPEX	Capital expenditure.
Casing	Metal pipe inserted into a well bore and cemented in place to protect both subsurface formations (such as groundwater) and the well bore. A surface casing is set first to protect groundwater. The production casing is the last one set. The production tubing (through which hydrocarbons flow to the surface) will be suspended inside the production casing.
Casing String	The steel tubing that lines a well after it has been drilled. It is formed from sections of steel tube screwed together.
Christmas tree	The assembly of valves, pipes, and fittings used to control the flow of oil and gas from a well.
Compressor	An equipment used to increase the pressure of natural gas so that it will flow more easily through a pipeline
Condensate	A term used to describe light liquid hydrocarbons separated from crude oil after production and sold separately. Hydrocarbons, which are in the gaseous state under reservoir conditions and which become liquid when temperature or pressure is reduced. A mixture of pentanes and higher hydrocarbons.

Cracking	Refinery process whereby large, heavy, complex hydro-carbon molecules are broken down into simpler and lighter molecules in order to derive a variety of fuel products.
Crude oil	A mixture of hydrocarbons that exists in liquid phase in natural underground reservoirs and remains liquid at atmospheric pressure after passing through surface separating facilities. Crude oil is refined to produce a wide array of petroleum products, including heating oils; gasoline, diesel and jet fuels; lubricants; asphalt; ethane, propane, and butane; and many other products used for their energy or chemical content.
Completion	The installation of permanent wellhead equipment for the production of oil and gas.
Connate Water	Salt water occurring with oil and gas in the reservoir.
Crane Barge	A large barge, capable of lifting heavy equipment onto offshore platforms. Also known as a "derrick barge".
Development well	A well drilled within the proved area of an oil or gas reservoir to the depth of a stratigraphic horizon known to be productive; a well drilled in a proven field for the purpose of completing the desired spacing pattern of production.
Development Phase	The phase in which a proven oil or gas field is brought into production by drilling production (development) wells.
Distillation	The first stage in the refining process in which crude oil is heated and unfinished petroleum products are initially separated.
Down hole	A term used to describe tools, equipment, and instruments used in the well bore, or conditions or techniques applying to the well bore.
Dry hole	An exploratory or development well found to be incapable of producing either oil or gas in sufficient quantities to justify completion as an oil or gas well.
Downstream	The oil industry term used to refer to all petroleum activities from the processing of refining crude oil into petroleum products to the distribution, marketing, and shipping of the products. The opposite of downstream is upstream.
Dry natural gas	Natural gas which remains after: (1) the liquefiable hydrocarbon portion has been removed from the gas

stream (i.e., gas after lease, field, and/or plant separation); and (2) any volumes of non-hydrocarbon gases have been removed where they occur in sufficient quantity to render the gas unmarketable. Dry natural gas is also known as consumer-grade natural gas.

Dry Gas	Natural gas composed mainly of methane with only minor amounts of ethane, propane and butane and little or no heavier hydrocarbons in the gasoline range.
E&P	Abbreviation for exploration and production. The "upstream" sector of the oil and gas industry.
Economy of scale	The principle that larger production facilities have lower unit costs than smaller facilities.
Equity crude oil	The proportion of production that a concession owner has the legal and contractual right to retain.
Enhanced Oil Recovery (EOR)	A process whereby oil is recovered other than by the natural pressure in a reservoir. This refers to a variety of processes to increase the amount of oil removed from a reservoir, typically by injecting a liquid (e.g., water, surfactant) or gas (e.g., nitrogen, carbon dioxide).
Exploration Drilling	Drilling carried out to determine whether hydrocarbons are present in a particular area or structure. Drilling done in search of new mineral deposits, on extensions of known ore deposits, or at the location of a discovery up to the time when the company decides that sufficient ore reserves are present to justify commercial exploration; Assessment drilling is reported as exploration drilling.
Exploration Phase	The phase of operations, which covers the search for oil or gas by carrying out detailed geological and geophysical surveys followed up where appropriate by exploratory drilling.
Exploratory well	A hole drilled: (a) to find and produce oil or gas in an area previously considered unproductive area; (b) to find a new reservoir in a known field, i.e., one previously producing oil and gas from another reservoir, or (c) to extend the limit of a known oil or gas reservoir. A well drilled in an unproven area. Also known as a "wildcat well".
Farm out (in) arrangement	An arrangement, used primarily in the oil and gas industry, in which the owner or lessee of mineral rights (the first party) assigns a working interest to an operator (the second party), the consideration for which is specified

exploration and/or development activities. The first party retains an overriding royalty or other type of economic interest in the mineral production. The arrangement from the viewpoint of the second party is termed a "farm-in arrangement."

f.a.s. value   Free alongside ship value. The value of a commodity at the port of exportation, generally including the purchase price plus all charges incurred in placing the commodity alongside the carrier at the port of exportation in the country of exportation.

Free on board (f.o.b.)   A sales transaction in which the seller makes the product available for pick up at a specified port or terminal at a specified price and the buyer pays for the subsequent transportation and insurance.

f.o.b. price   The price actually charged at the producing country's port of loading. The reported price should be after deducting any rebates and discounts or adding premiums where applicable and should be the actual price paid with no adjustment for credit terms.

Field   An area consisting of a single reservoir or multiple reservoirs all grouped on, or related to, the same individual geological structural feature and/or stratigraphic condition. There may be two or more reservoirs in a field that are separated vertically by intervening impervious strata or laterally by local geologic barriers, or by both.

Flare   Gas disposed of by burning in flares usually at the production sites or at gas processing plants.

Fracturing   The application of hydraulic pressure to the reservoir formation to create fractures through which oil or gas may move to the well bore.

Fossil fuel   An energy source formed in the earths crust from decayed organic material. The common fossil fuels are petroleum, coal, and natural gas.

Fractionation   The process by which saturated hydrocarbons are removed from natural gas and separated into distinct products, or "fractions," such as propane, butane, and ethane.

Fishing   Retrieving objects from the borehole, such as a broken drill string or tools.

Formation Water   Salt water underlying gas and oil in the formation.

Fracturing   A method of breaking down a formation by pumping fluid at very high pressures. The objective is to increase production rates from a reservoir.

Gas   A non-solid, non-liquid combustible energy source that includes natural gas, coke-oven gas, blast-furnace gas, and refinery gas.

Gas Field   A field containing natural gas but no oil.

Gas well   A well completed for production of natural gas from one or more gas zones or reservoirs. Such wells contain no completions for the production of crude oil.

Gas Injection   The process whereby separated associated gas is pumped back into a reservoir for conservation purposes or to maintain the reservoir pressure.

Gas/Oil Ratio   The volume of gas at atmospheric pressure produced per unit of oil produced.

Gas oil   A medium-distilled oil from the refining process. Often used in diesel fuel. European and Asian designation for No. 2 heating oil and No. 2 diesel fuel (in USA).

Gallon   A volumetric measure equal to 4 quarts (231 cubic inches) used to measure fuel oil. One barrel equals 42 gallons.

Geological and Geophysical (G&G) costs   Costs incurred in making geological and geophysical studies, including, but not limited to, costs incurred for salaries, equipment, obtaining rights of access, and supplies for scouts, geologists, and geophysical crews.

Hydrocarbon   An organic compound of hydrogen and carbon, called petroleum. The molecular structure of the hydrocarbon varies from the simplest, methane (CH_4, a constituent of natural gas), to the very heavy and very complex. Octane, a constituent of crude oil, is one of the heavier, more complex molecules (C8H18).

Heating value   The average number of British thermal units per cubic foot of natural gas as determined from tests of fuel samples.

Heavy gas   Oil Petroleum distillates with an approximate boiling range from 651 degrees Fahrenheit to 1000 degrees Fahrenheit.

Heavy oil   The fuel oils remaining after the lighter oils have been distilled off during the refining process. Except for start-up and flame stabilization, virtually all petroleum used in steam plants is heavy oil. Includes fuel oil numbers 4, 5, and 6; crude; and topped crude.

Henry Hub   A pipeline hub on the Louisiana Gulf coast. It is the delivery point for the natural gas futures contract on the New York Mercantile Exchange (NYMEX).

Hydraulic fracturing   Fracturing of rock at depth with fluid pressure. Hydraulic fracturing at depth may be accomplished by pumping water into a well at very high pressures. Under natural conditions, vapor pressure may raise high enough to cause fracturing in a process known as hydrothermal brecciation.

Hydro chlorofluoro-carbons (HCFCs)   Chemicals composed of one or more carbon atoms and varying numbers of hydrogen, chlorine, and fluorine atoms.

Hydro fluoro-carbons (HFCs)   A group of man-made chemicals composed of one or two carbon atoms and varying numbers of hydrogen and fluorine atoms. Most HFCs have 100 year Global Warming Potentials in the thousands.

Improved recovery   Extraction of crude oil or natural gas by any method other than those that rely primarily on natural reservoir pressure, gas lift, or a system of pumps.

Integrated E&P company   When applied to an oil company, it indicates a firm that operates in both the upstream and downstream sectors (from exploration through refining and marketing).

Injection Well   A well used for pumping water or gas into the reservoir.

Jacket   The lower section or "legs" of an offshore platform.

Kick   A well is said to "kick" if the formation pressure exceeds the pressure exerted by the mud column.

Lay Barge   A barge that is specially equipped to lay submarine pipelines.

Liquefied Natural Gas (LNG)   Oilfield or naturally occurring gas, chiefly methane, liquefied for transportation. Natural gas (primarily methane) that has been liquefied by reducing its temperature to –260 degrees Fahrenheit at atmospheric pressure.

Liquefied Petroleum Gas (LPG)   Light hydrocarbon material, gaseous at atmospheric temperature and pressure, held in the liquid state by pressure to facilitate storage, transport and handling. Commercial liquefied gas consists essentially of either propane or butane, or mixtures thereof.

Light gas oils   Liquid petroleum distillates heavier than naphtha, with an approximate boiling range from 401 degrees to 650 degrees Fahrenheit. Light oil Lighter fuel oils distilled off during the refining process. Virtually all petroleum used in internal combustion and gas-turbine engines is light oil. Includes fuel oil numbers 1 and 2, kerosene, and jet fuel.

Lease   A legal document conveying the right to drill for oil and gas, or the tract of land on which a lease has been obtained where the producing wells and production equipment are located.

Lifting costs   The cost of producing oil from a well or lease.

Log   To conduct a survey inside a borehole to gather information about the subsurface formations; the results of such a survey. Logs typically consist of several curves on a long grid that describe properties within the well bore or surrounding formations that can be interpreted to provide information about the location of oil, gas, and water. Also called well logs, borehole logs, wireline logs.

Multiple completion well   A well equipped to produce oil and/or gas separately from more than one reservoir. Such wells contain multiple strings of tubing or other equipment that permit production from the various completions to be measured and accounted for separately. For statistical purposes, a multiple completion well is reported as one well and classified as either an oil well or a gas well.

Midstream   A term sometimes used to refer to those industry activities that fall between exploration and production (upstream) and refining and marketing (downstream). The term is most often applied to pipeline transportation of crude oil and natural gas.

Mud   A mixture of base substance and additives used to lubricate the drill bit and to counteract the natural pressure of the formation.

Natural reservoir pressure   The energy within an oil or gas reservoir that causes the oil or gas to rise (unassisted by other forces) to the earth's surface when the reservoir is penetrated by oil or gas well. The energy may be the result of "dissolved gas drive," "gas cap drive," or "water drive." Regardless of the type of drive, the principle is the same: the energy of the gas or water, creating a natural pressure, forces the oil or gas to the well bore.

Natural Gas	Gas, occurring naturally and often found in association with crude petroleum. A gaseous mixture of hydrocarbon compounds, the primary one being methane.
Natural gas hydrates	Solid, crystalline, wax-like substances composed of water, methane, and usually a small amount of other gases, with the gases being trapped in the interstices of a water-ice lattice. They form beneath permafrost and on the ocean floor under conditions of moderately high pressure and at temperatures near the freezing point of water.
Natural Gas Liquids (NGL)	The portions of gas from a reservoir that are liquefied at the surface in separators, field facilities, or gas processing plants. NGL from gas processing plants is also called Liquefied Petroleum Gas (LPG).Natural gas liquid. Liquid hydrocarbons found in association with natural gas.
Non-associated gas	Natural gas produced from a reservoir that does not contain significant quantities of crude oil.
Natural Gasoline	A liquid similar to motor fuel recovered as drip gasoline or produced in natural gasoline plants, but generally having a lower octane number and being more volatile than commercial motor fuel.
Natural Gas Storage	Use of a depleted formation (or well) nears a market to store gas bought in from another field or location.
Offshore	That geographic area that lies seaward of the coastline. In general, the coastline is the line of ordinary low water along with that portion of the coast that is in direct contact with the open sea or the line marking the seaward limit of inland water.
Oil, crude	Petroleum oil and other hydrocarbons regardless of gravity which are produced at the wellhead in liquid form and the liquid hydrocarbons known as distillate or condensate recovered or extracted from gas.
Oil In Place	An estimated measure of the total amount of oil contained in a reservoir, and, as such, a higher figure than the estimated recoverable reserves of oil.
Oil reservoir	An underground pool of liquid consisting of hydrocarbons, sulfur, oxygen, and nitrogen trapped within a geological

formation and protected from evaporation by the overlying mineral strata.

Oil well   A well completed for the production of crude oil from at least one oil zone or reservoir.

OPEX   Operating expenditure.

Operator   Term used to describe a company appointed by venture stake holders to take primary responsibility for day-to-day operations for a specific plant or activity.

Pay Zone   Rock in which oil and gas are found in exploitable quantities.

Permeability   The property of a formation, which quantifies the flow of a fluid through the pore spaces and into the well bore.

Petroleum   A generic name for hydrocarbons, including crude oil, natural gas liquids, natural gas and their products.

Platform   An offshore structure that is permanently fixed to the seabed.

Petroleum   A complex liquid mixture of hydrocarbons, oily and inflammable in character.

Petrochemicals   Chemicals such as ethylene, propylene and benzene that are derived from petroleum.

Play   A play is a set of known or postulated oil and (or) gas accumulations sharing similar geologic, geographic, and temporal properties, such as source rock, migration pathway, timing, trapping mechanism, and hydrocarbon type.

Plug   To stop the flow of oil from one stratum to another in connection with the abandoning of a well.

Pour point   The ability of crude oil to flow at low temperatures.

Porosity   The percentage of void in a porous rock compared to the solid formation. A ratio between the volume of the pore space in reservoir rock and the total bulk volume of the rock. The pore space determines the amount of space available for storage of fluids.

Permeability   A measure of the ability of a rock to transmit fluid through pore spaces.

Produced water   The water extracted from the subsurface with oil and gas. It may include water from the reservoir, water that has been injected into the formation and any chemicals added during the production/treatment process.

Primary energy	All energy consumed by end users, excluding electricity but including the energy consumed at electric utilities to generate electricity.
Primary energy consumption	Primary energy consumption is the amount of site consumption, plus losses that occur in the generation, transmission, and distribution of energy.
Probable energy reserves	Estimated quantities of energy sources that, on the basis of geologic evidence that supports projections from proved reserves can reasonably be expected to exist and be recoverable under existing economic and operating conditions.
Producer	A company engaged in the production and sale of natural gas from gas or oil wells with delivery generally at a point at or near the wellhead, the field, or the tailgate of a gas processing plant.
Producing property	A term often used in reference to a property, well, or mine that produces wasting natural resources. The term means a property that produces in paying quantities (that is, one for which proceeds from production exceed operating expenses).
Prospecting	The search for an area of probable mineralization; the search normally includes topographical, geological, and geophysical studies of relatively large areas undertaken in an attempt to locate specific areas warranting detailed exploration.
Prospecting costs	Direct and indirect costs incurred to identify areas of interest that may warrant detailed exploration.
Possible Reserves	Those reserves which at present cannot be regarded as 'probable' but are estimated to have a significant but less than 50% chance of being technically and economically producible.
Proved energy reserves	Estimated quantities of energy sources that analysis of geologic and engineering data demonstrates with reasonable certainty are recoverable under existing economic and operating conditions. The location, quantity, and grade of the energy source are usually considered to be well established in such reserves.
Primary Recovery	Recovery of oil or gas from a reservoir purely by using the natural pressure in the reservoir to force the oil or gas out.

Probable
Reserves
Those reserves which are not yet proven but which are estimated to have a better than 50% chance of being technically and economically producible.

Proven Field
An oil and/or gas field whose physical extent and estimated reserves have been determined.

Proven Reserves
Those reserves which on the available evidence are virtually certain to be technically and economically producible (i.e. having a better than 90% chance of being produced).

Recoverable
Reserves
That proportion of the oil and/gas in a reservoir that can be removed using currently available techniques.

Recoverable
proved reserves
The proved reserves of natural gas as of December 31 of any given year are the estimated quantities of natural gas which geological and engineering data demonstrates with reasonable certainty to be recoverable in the future from known natural oil and gas reservoirs under existing economic and operating conditions.

Recovery Factor
The ratio of recoverable oil and/or gas reserves to the estimated oil and/or gas in place in the reservoir.

Reserve
That portion of the demonstrated reserve base that is estimated to be recoverable at the time of determination. The reserve is derived by applying a recovery factor to that component of the identified coal resource designated as the demonstrated reserve base.

Reserves, net
Includes all proved reserves associated with the company's net working interests.

Reservoir
The underground formation where oil and gas has accumulated It consists of a porous rock to hold the oil or gas, and a cap rock that prevents its escape.A porous and permeable underground formation containing an individual and separate natural accumulation of producible hydrocarbons (crude oil and/or natural gas) which is confined by impermeable rock or water barriers and is characterized by a single natural pressure system.

Reserve revisions
Changes to prior year-end proved reserves estimates, either positive or negative, resulting from new information other than an increase in proved acreage (extension). Revisions include increases of proved reserves associated with the installation of improved recovery techniques or

equipment. They also include correction of prior year arithmetical or clerical errors and adjustments to prior year-end production volumes to the extent that these alter reserves estimates.

Reserve additions The estimated original, recoverable, salable, and new proved reserves credited to new fields, new reservoirs, new gas purchase contracts, amendments to old gas purchase contracts, or purchase of gas reserves in-place that occurred during the year and had not been previously reported. Reserve additions refer to domestic in-the-ground natural gas reserve additions and do not refer to interstate pipeline purchase agreements; contracts with foreign suppliers; coal gas, SNG, or LNG purchase arrangements.

Reserves changes Positive and negative revisions, extensions, new reservoir discoveries in old fields, and new field discoveries that occurred during the report year.

Royalty A percentage interest in the value of production from a lease that is retained and paid to the mineral rights owner.

Riser (Drilling) A pipe between a seabed BOP and a floating drilling rig.

Riser (Production) The section of pipe work that joins a seabed wellhead to the Christmas tree.

Roughneck Drill crewmembers who work on the derrick floor, screwing together the sections of drill pipe when running or pulling a drill string.

Roustabout Drill crewmembers who handle the loading and unloading of equipment and assist in general operations around the rig.

Secondary Recovery Recovery of oil or gas from a reservoir by artificially maintaining or enhancing the reservoir pressure by injecting gas, water or other substances into the reservoir rock.

Shutdown A production hiatus during which the platform ceases to produce while essential maintenance work is undertaken.

Spud-In The operation of drilling the first part of a new well.

Suspended Well A well that has been capped off temporarily.

Separation The process of separating liquid and gas hydrocarbons and water. This is typically accomplished in a pressure vessel

	at the surface, but newer technologies allow separation to occur in the well bore under certain conditions.
Service well	A well drilled, completed, or converted for the purpose of supporting production in an existing field. Wells of this class also are drilled or converted for the following specific purposes: gas injection (natural gas, propane, butane or fuel-gas); water injection; steam injection; air injection; salt water disposal; water supply for injection; observation; and injection for in-situ combustion.
Shell storage capacity	The design capacity of a petroleum storage tank which is always greater than or equal to working storage capacity.
Short ton	A unit of weight equal to 2,000 pounds.
Shrinkage	The volume of natural gas that is transformed into liquid products during processing, primarily at natural gas liquids processing plants.
Sidetrack drilling	This is a remedial operation that results in the creation of a new section of well bore for the purpose of (1) detouring around junk, (2) redrilling lost holes, or (3) straightening key seats and crooked holes.
Sour crude oil	Oil containing free sulfur or other sulfur compounds whose total sulfur content is in excess of 1 percent.
Sour gas	natural gas containing hydrogen sulfide.
Specific gravity	A measure of the density of a material usually obtained by comparing it with water.
Stimulation	The term used for several processes to enlarge old channels, or create new ones, in the producing formation of a well designed to enhance production. Examples include acidizing and fracturing.
Strategic Petroleum Reserve (SPR)	Petroleum stocks planned to be maintained by the Central Government for use during periods of major supply interruption.
Stripper well	An oil or gas well that produces at relatively low rates. For oil, stripper production is usually defined as production rates of between 5 and 15 barrels of oil per day. Stripper gas production would generally be anything less than 60 thousand cubic feet per day.
Sweet crude oil	Crude oil with low sulphur content.

Sulphur
A yellowish nonmetallic element, sometimes known as "brimstone." It is present at various levels of concentration in many fossil fuels whose combustion releases sulfur compounds that are considered harmful to the environment. Some of the most commonly used fossil fuels are categorized according to their sulfur content, with lower sulfur fuels usually selling at a higher price.

Supplemental gas
Any gaseous substance introduced into or commingled with natural gas that increased the volume available for disposition. Such substances include, but are not limited to, propane-air, refinery gas, coke-oven gas, still gas, manufactured gas, biomass gas, or air or inerts added for Btu stabilization.

Supply source
May be a single completion, a single well, a single field with one or more reservoirs, several fields under a single gas-purchase contract, miscellaneous fields, a processing plant, or a field area.

Tank farm
An installation used by trunk and gathering pipeline companies, crude oil producers, and terminal operators (except refineries) to store crude oil.

Tanker and barge
Vessels that transport crude oil or petroleum products.

Tariff
A published volume of rate schedules and general terms and conditions under which a product or service will be supplied.

Throughput
The total amount of raw materials processed by a refinery or other plant in a given period.

Tool Pusher
Second-in-command of a drilling crew under the drilling superintendent. Responsible for the day-to-day running of the rig and for ensuring that all the necessary equipment are available.

Topsides
The superstructure of a platform.

Total discoveries
The sum of extensions, new reservoir discoveries in old fields, and new field discoveries that occurred during the report year.

Total liquid hydrocarbon reserves
The sum of crude oil and natural gas liquids reserves volumes.

Trap	The occurrence of those structures, pinch-outs, permeability changes, and similar features necessary for the entrapment of oil and (or) gas in at least one accumulation of the minimum size. Included in this attribute is existence of seals sufficient for entrapping hydrocarbons and capable of holding oil and gas accumulations during appropriate ranges of geologic time.
Terminal	Plant and equipment designed to receive and process crude oil or gas to remove water and impurities.
Transport	Movement of natural, synthetic, and/or supplemental gas between points beyond the immediate vicinity of the field or plant from which produced except (1) for movements through well or field lines to a central point for delivery to a pipeline or processing plant within the same state or (2) movements from a city gate point of receipt to consumers through distribution mains.
Transshipment	A method of ocean transportation whereby ships off-load their oil cargo to a deepwater terminal, floating storage facility, temporary storage, or to one or more smaller tankers from which or in which the oil is then transported to a market destination.
Trunk line	A main pipeline.
Upstream	The processes of exploring for oil; developing oil fields; and producing oil from the oil fields; in other words, the exploration and production portions of the oil and gas industry. The opposite of upstream is downstream.
Underground gas storage	The use of sub-surface facilities for storing gas that has been transferred from its original location. The facilities are usually hollowed-out salt domes, geological reservoirs (depleted oil or gas fields) or water-bearing sands topped by an impermeable cap rock (aquifer).
Underground injection	The placement of gases or fluids into an underground reservoir through a well bore. May be used as part of enhanced oil recovery or water flooding processes or for disposal of produced water.
Undiscovered recoverable reserves (crude oil and natural gas)	Those economic resources of crude oil and natural gas, yet undiscovered, that are estimated to exist in favorable geologic settings.

Unit value, wellhead	The wellhead sales price, including charges for natural gas plant liquids subsequently removed from the gas; gathering and compression charges; and state production, severance, and/or similar charges.
Vessel	A ship used to transport crude oil, petroleum products, or natural gas products. Vessel categories are as follows: Ultra Large Crude Carrier (ULCC), Very Large Crude Carrier (VLCC), Other Tanker, and Specialty Ship (LPG/LNG).
Viscosity	One of the physical properties of a liquid, namely, its ability to flow. It is expressed inversely, i.e. the less viscous the fluid the greater its mobility. The viscosity of oil in a reservoir affects the rate and amount of recovery. While viscosity is related to specific gravity, it is also affected by the amount of gas in solution in the oil. Greater recoveries can be obtained where the solution gas is not allowed to escape prior to the time the oil is removed from the reservoir.
Well	A hole drilled in the earth for the purpose of (1) finding or producing crude oil or natural gas; or (2) producing services related to the production of crude or natural gas. crude oil or natural gas; or (2) producing services related to the production of crude or natural gas.
Wellhead	The point at which the crude (and/or natural gas) exits the ground. Following historical precedent, the volume and price for crude oil production are labeled as "wellhead," even though the cost and volume are now generally measured at the lease boundary. In the context of domestic crude price data, the term "wellhead" is the generic term used to reference the production site or lease property.
Well Log	A record of geological formation penetrated during drilling, including technical details of the operation.
Well servicing	Maintenance work performed on an oil or gas well to improve or maintain the production.
Wildcat Well	A well drilled in an unproven area. Also known as an "exploration well". [The term comes from exploration wells in West Texas in the 1920s. Wildcats were abundant in the locality, and those unlucky enough to be shot were hung from oil derricks.] a well drilled in an area where no current oil or gas production exists.

Wet natural gas   A mixture of hydrocarbon compounds and small quantities of various nonhydrocarbons existing in the gaseous phase or in solution with crude oil in porous rock formations at reservoir conditions. Under reservoir conditions, natural gas and its associated liquefiable portions occur either in a single gaseous phase in the reservoir or in solution with crude oil and are not distinguishable at the time as separate substances.

Work over   Operations on a producing well to restore or increase production. A work over may be performed to stimulate the well, remove sand or wax from the well bore, to mechanically repair the well, or for other reasons.

WTI   West Texas Intermediate, a type of crude oil commonly used as a price benchmark.